15

D1112675

COPERNICAN

The

REVOLUTION

THOMAS S. KUHN

3.68

MJF BOOKS

NEW YORK

Published by MJF Books
Fine Communications
Two Lincoln Square
60 West 66th Street
New York, NY 10023

ISBN 1-56731-217-9

This edition published by arrangement with Harvard University Press.

Manufactured in the United States of America on acid-free paper

MJF Books and the MJF colophon are trademarks of Fine Creative Media, Inc.

10   9   8   7   6   5   4   3   2   1

TO L. K. NASH

FOR A VEHEMENT COLLABORATION

PREFACE

The story of the Copernican Revolution has been told many times before, but never, to my knowledge, with quite the scope and object aimed at here. Though the Revolution's name is singular, the event was plural. Its core was a transformation of mathematical astronomy, but it embraced conceptual changes in cosmology, physics, philosophy, and religion as well. These individual aspects of the Revolution have been examined repeatedly, and without the resulting studies this book could not have been written. The Revolution's plurality transcends the competence of the individual scholar working from primary sources. But both specialized studies and the elementary works patterned on them necessarily miss one of the Revolution's most essential and fascinating characteristics — a characteristic which arises from the Revolution's plurality itself.

Because of its plurality, the Copernican Revolution offers an ideal opportunity to discover how and with what effect the concepts of many different fields are woven into a single fabric of thought. Copernicus himself was a specialist, a mathematical astronomer concerned to correct the esoteric techniques used in computing tables of planetary position. But the direction of his research was often determined by developments quite foreign to astronomy. Among these were medieval changes in the analysis of falling stones, the Renaissance revival of an ancient mystical philosophy which saw the sun as the image of God, and the Atlantic voyages which widened the terrestrial horizons of Renaissance man. Even stronger filiations between distinct fields of thought appear in the period after the publication of Copernicus' work. Though his *De Revolutionibus* consists principally of mathematical formulas, tables, and diagrams, it could only be assimilated by men able to create a new physics, a new conception of space, and a new idea of man's relation to God. Creative interdisciplinary ties like these play many and varied roles in the Copernican Revolution. Specialized accounts are inhibited both by aim and method from examining the nature of these ties and their effects upon the growth of human knowledge.

This account of the Copernican Revolution therefore aims to display the significance of the Revolution's plurality, and that object is probably the book's most important novelty. Pursuit of the object has, however, necessitated a second innovation. This book repeatedly violates the institutionalized boundaries which separate the audience for "science" from the audience for "history" or "philosophy." Occasionally it may seem to be two books, one dealing with science, the other with intellectual history.

The combination of science and intellectual history is, however, essential in approaching the plural structure of the Copernican Revolution. The Revolution centered in astronomy. Neither its nature, its timing, nor its causes can be understood without a firm grasp upon the data and concepts that were the tools of planetary astronomers. Astronomical observations and theories therefore make up the essential "scientific" component which dominates my first two chapters and recurs throughout the remainder of this book. They do not, however, make up the whole book. Planetary astronomy was never a totally independent pursuit with its own immutable standards of accuracy, adequacy, and proof. Astronomers were trained in other sciences as well, and they were committed to various philosophical and religious systems. Many of their nonastronomical beliefs were fundamental first in postponing and then in shaping the Copernican Revolution. These nonastronomical beliefs compose my "intellectual history" component, which, after Chapter 2, parallels the scientific. Given the purpose of this book, the two are equally fundamental.

Besides, I am not convinced that the two components are really distinct. Except in occasional monographs the combination of science and intellectual history is an unusual one. Initially it may therefore seem incongruous. But there can be no intrinsic incongruity. Scientific concepts are ideas, and as such they are the subject of intellectual history. They have seldom been treated that way, but only because few historians have had the technical training to deal with scientific source materials. I am myself quite certain that the techniques developed by historians of ideas can produce a kind of understanding that science will receive in no other way. Though no elementary book can fully document that thesis, this one should provide at least preliminary evidence.

Indeed it has already provided some evidence. The book grows out of a series of lectures delivered each year since 1949 in one of the science General Education courses at Harvard College, and in that application the combination of technical and intellectual-historical materials has been quite successful. Since students in this General Education course do not intend to continue the study of science, the technical facts and theories that they learn function principally as paradigms rather than as intrinsically useful bits of information. Furthermore, though the technical scientific materials are essential, they scarcely begin to function until placed in a historical or philosophical framework where they illuminate the way in which science develops, the nature of science's authority, and the manner in which science affects human life. Once placed in that framework, however, the Copernican system or any other scientific theory has relevance and appeal for an audience far broader than either the scientific or the undergraduate community. Though my first purpose in writing it was to supply reading for the Harvard course and for others like it, this book, which is not a text, is also addressed to the general reader.

Many friends and colleagues, by their advice and criticism, have helped to shape this book, but none has left so large or significant a mark as Ambassador James B. Conant. Work with him first persuaded me that historical study could yield a new sort of understanding of the structure and function of scientific research. Without my own Copernican revolution, which he fathered, neither this book nor my other essays in the history of science would have been written.

Mr. Conant also read the manuscript, and its early chapters show many signs of his productive criticisms. Others who will recognize here and there the effects of their useful suggestions include Marie Boas, I. B. Cohen, M. P. Gilmore, Roger Hahn, G. J. Holton, E. C. Kemble, P. E. LeCorbeiller, L. K. Nash, and F. G. Watson. Each has applied critical talent to at least one chapter; several read the entire manuscript in an earlier version; and all have rescued me from mistakes or ambiguities. The advice of Mason Hammond and Mortimer Chambers has given my occasional Latin translations an assurance that they would not otherwise possess. Arnolfo Ferruolo first introduced me to Ficino's *De Sole* and showed me that Copernicus' attitude toward the sun is an integral part of a Renaissance tradition more striking even in the arts and literature than in the sciences.

The illustrations display the skill, but scarcely the patience, with which Miss Polly Horan has translated and retranslated my vague directions into communicative symbols. J. D. Elder and the staff of the Harvard University Press have given me constant and sympathetic guidance in the arduous transmutation to type of a manuscript that conforms neither to the rules for scientific publication nor to those for history. The index attests the industry and intelligence of W. J. Charles.

The joint generosity of Harvard University and the John Simon Guggenheim Memorial Foundation provided the year's leave of absence during which most of my manuscript was first prepared. I am also grateful to the University of California for a small grant which assisted in the final preparation of the manuscript and in seeing it through the press.

My wife has been an active participant throughout the book's development, but that participation is the least of her contributions to it. Brain children, particularly someone else's, are the most obstreperous members of any household. Without her continuing toleration and forebearance this one would never have survived.

T. S. K.

Berkeley, California
November 1956

Note to the Seventh Printing. This printing contains a number of corrections and textual changes inadvertently omitted from earlier Harvard editions. With this and subsequent printings, all changes previously introduced in the Random House and Vintage paperback editions are also included in the Harvard Paperback together with a few minor corrections after the earlier paperback editions were prepared.

CONTENTS

FOREWORD

In Europe west of the Iron Curtain, the literary tradition in education still prevails. An educated man or woman is a person who has acquired a mastery of several tongues and retained a working knowledge of the art and literature of Europe. By a working knowledge I do not refer to a scholarly command of the ancient and modern classics or a sensitive critical judgment of style or form; rather, I have in mind a knowledge which can be readily worked into a conversation at a suitable social gathering. An education based on a carefully circumscribed literary tradition has several obvious advantages: the distinction between the 5 to 10 percent of the population who are thus educated and the others makes itself evident almost automatically when ladies and gentlemen converse. For those who truly enjoy art, literature and music, there is a comforting sense of solidarity. For others who feel compelled to enter into a discussion of these subjects, the area of maneuver is conveniently limited; not too much effort is required to keep fresh a portion of the knowledge painfully acquired at school. The price of admission to the cultural tradition of a European nation is paid once and for all when one is young. Theoretically, this price is eight or nine years of hard work in special schools whose curricula are centered on the languages and literature of Greece and Rome. I say theoretically, since in practice the study of modern languages has in this century made inroads on the study of Greek and to some extent on a knowledge of Latin as well. But even these changes have not fundamentally altered the basic idea of education for the few as being the consequence of long years of school work devoted to the study of languages and the literature of Europe.

There have been attacks on this type of education off and on for a century at least. The claims of the physical sciences for a greater share of the curriculum have been pressed and such claims have usually been associated with demands for the substitution of modern languages for the ancient. The place of mathematics has hardly been at issue, since a thorough study of mathematics, including the calculus, has long been accepted as a matter of course in all the curricula of those special schools which prepare a student to enter the university. Several generations ago, as a clear-cut alternative to the classical curriculum, a course of study was suggested which would be based on physics, chemistry, mathematics, and modern languages. But the proponents of the classical course are still vigorous and effective. In Germany, at least, a series of compromises seems to have been the result of the argument. But because of the importance attached to a study of languages, it is hardly too much to say that the literary tradition still dominates. Even in those schools which devote the most time to science, it would hardly be correct to say that the scientific tradition had replaced the literary. Rather, one might say, in varying degrees German students entering a university have acquired a considerable amount of information about the physical sciences. But whether such knowledge subsequently affects the attitude of those who do not proceed with a scientific education is at least an open question. There seems little or no concern with changing educational methods so that the nonscientist will acquire a better understanding of science. Indeed, it would not be strange if those whose education was primarily literary would question whether understanding science was a matter of importance to anybody but scientists or engineers.

In the United States the European literary tradition as a basis for education disappeared almost a hundred years ago, or rather was transformed beyond recognition. But it has not been replaced by an education based on the physical sciences, mathematics, and modern languages. Some would say there has simply been no replacement. At all events, there have been repeated attempts to provide some broad base for the cultural life of the nation — broad enough to include the physical, biological, and social sciences as well as the Anglo-Saxon literary tradition and concern with art forms from various civilizations. Whether such attempts, directed toward producing a future citizen of a democracy who will be an enthusiastic participant in the nation's developing culture, have created a medium sufficiently nutrient for the life of the spirit in America may be a question. But no one can deny that those responsible for the attempts have, with few exceptions, endeavored to find a suitable place for the scientific tradition.

Experience has shown, however, both in the United States and in the modern schools of Europe, how difficult it is to place the study of science on anything like the same footing as the study of literature or art or music. A scientist or engineer may be able to participate in a stimulating manner in a discussion of pictures or books or plays, but it is very hard indeed to keep a conversation going about physical science in which the majority of the participants are not themselves scientists or engineers. (And while I should be the first to deny that facility in conversation was a goal of education, nevertheless listening in on a social gathering may be a permissible diagnostic method.)

It is quite clear that studying science and studying literature in school or college do not leave the same sort of residue in the student's mind. A knowledge of the chemistry of metals and a knowledge of Shakespeare's plays are two entirely different kinds of knowledge as far as the needs of a human being are concerned. Of course, it is not necessary to pick an example from the natural sciences; for the "chemistry of metals" in the preceding sentence one could quite as well substitute the words "Latin grammar." Expressed in very simple terms, the difference lies in the fact that Shakespeare's plays have been and still are the subjects of endless debates in which the style and the characters have been criticized from every conceivable angle, and strong words of admiration and condemnation are constantly to be heard. No one either admires or condemns the metals or the behavior of their salts.

No, something more than a study of science as a body of organized knowledge, something more than an understanding of scientific theories is required to make educated people ready to accept the scientific tradition alongside that literary tradition which still underlies even the culture of the United States. This is so because the difficulties of assimilating science into Western culture have increased with the centuries. When in the time of Louis XIV scientific academies were formed, new discoveries and new theories in science were far more accessible to educated persons than today; the situation was the same as late as the Napoleonic Wars. Sir Humphrey Davy fascinated London society at the beginning of the 19th century by his lectures on chemistry illustrated by spectacular experiments. Fifty years later Michael Faraday delighted audiences of young and old who came to hear him in the auditorium of the Royal Institution in London; his lectures on the chemistry of the candle are classic examples of popularizing science. In our own times there has been no lack of attempts along similar lines; but the obstacles to be overcome have grown with the years. Spectacular lecture-table experiments no longer astonish and please sophisticated audiences as they once did; large-scale engineering outplays them almost daily. The scientific novelties of the current year are too numerous as well as too complicated to form a topic of conversation among laymen. The advances are made so rapidly and on so many fronts that the layman is bewildered by the news; furthermore, to have any comprehension of the significance of a scientific break-through one needs to be well versed on the state of the science in question before the successful attack was launched. Even those trained in one branch of science find it difficult to understand what is going on in a distant field. For example, physicists are hardly in a position to read even summary papers written by geneticists for other geneticists, and vice versa. For the large group of people with scientific and engineering training who wish to keep abreast of the progress of science in general, there are some excellent periodicals, and useful books are published from time to time. But I doubt very much if this effort to popularize science reaches those who are not directly connected with the physical or biological sciences or their application. And some attempts at popularization are so superficial and sensational as to be of no value for the purposes of providing a basis for the understanding of science by nonscientists.

In the last ten or fifteen years there has been growing concern in American colleges as to the place of the physical and biological sciences in the curriculum. The orthodox first-year courses in physics, chemistry, and biology have been felt by many to be unsatisfactory for those students who do not intend to enter into an intensive study of science, engineering, or medicine. Various proposals have been made and various experiments tried involving new types of scientific courses which would be part of a liberal arts or general education program. In particular, more emphasis on the history of science has been recommended and in this recommendation I have heartily joined. Actually, experience with one type of historical approach in Harvard College for several years has increased my conviction as to the possibilities inherent in the study of the history of science, particularly if combined with an analysis of the various methods by which science has progressed. While recognizing the educational value of an over-all survey of the history of science in the last 300 years, I believe more benefit can be obtained by an intensive study of certain episodes in the development of physics, chemistry, and biology. This conviction has found expression in a series of pamphlets entitled "Harvard Case Histories in Experimental Science."

The cases considered in the Harvard series are relatively narrowly restricted in point of both chronology and subject matter. The aim has been to develop in the student some understanding of the interrelation between theory and experiment and some comprehension of the complicated train of reasoning which connects the testing of a hypothesis with the actual experimental results. To this end an original scientific paper is reprinted and forms the basis of the case; the reader is guided by comments of the editors to follow as far as possible the investigator's own line of reasoning. It is left to the professors who use these pamphlets to fit the case in question into a larger framework of the advance of science on a broad front.

The Harvard Case Histories are too limited in scope and too much concerned with experimental details and analysis of methods for the general reader. Furthermore, though the episodes chosen are all of them important in the history of physics, chemistry, and biology, their significance is not at once apparent to the uninitiated. The reader will soon be aware that the present volume does not suffer from these defects. Everyone knows of the impact on Western culture of the change from an Aristotelian universe centered on the earth to the Copernican universe. Professor Kuhn is concerned not with one event in the history of science but with a series of connected events influenced by and in turn influencing the attitude of learned men far removed in their interests from the field of astronomy itself. He has not set himself the relatively easy task of merely retelling the story of the development of astronomy during the revolutionary period. Rather he has succeeded in presenting an analysis of the relation between theory and observation and belief, and he has boldly faced such embarrassing questions as why brilliant, devoted, and completely sincere students of nature should have delayed so long in accepting the heliocentric arrangement of the planets. This book is no superficial account of the work of scientists; rather it is a thorough exposition of one phase of scientific work, from which the careful reader may learn about the curious interplay of hypothesis and experiment (or astronomical observation) which is the essence of modern science but largely unknown to the nonscientist.

It is not my purpose in this foreword, however, to attempt to present in capsule form a summary of the lessons on understanding science to be derived from reading what Professor Kuhn has written. Rather, I wish to register my conviction that the approach to science presented in this book is the approach needed to enable the scientific tradition to take its place alongside the literary tradition in the culture of the United States. Science has been an enterprise full of mistakes and errors as well as brilliant triumphs; science has been an undertaking carried out by very fallible and often highly emotional human beings; science is but one phase of the creative activities of the Western world which have given us art, literature, and music. The changes in man's views about the structure of the universe portrayed in the following pages affect to some degree the outlook of every educated person of our times; the subject matter is of deep significance in and by itself. But over and above the importance of this particular astronomical revolution, Professor Kuhn's handling of the subject merits attention, for, unless I am much mistaken, he points the way to the road which must be followed if science is to be assimilated into the culture of our times.

JAMES B. CONANT

THE
COPERNICAN REVOLUTION

1

THE ANCIENT

TWO-SPHERE UNIVERSE

Copernicus and the Modern Mind

The Copernican Revolution was a revolution in ideas, a transformation in man's conception of the universe and of his own relation to it. Again and again this episode in the history of Renaissance thought has been proclaimed an epochal turning point in the intellectual development of Western man. Yet the Revolution turned upon the most obscure and recondite minutiae of astronomical research. How can it have had such significance? What does the phrase "Copernican Revolution" mean?

In 1543, Nicholas Copernicus proposed to increase the accuracy and simplicity of astronomical theory by transferring to the sun many astronomical functions previously attributed to the earth. Before his proposal the earth had been the fixed center about which astronomers computed the motions of stars and planets. A century later the sun had, at least in astronomy, replaced the earth as the center of planetary motions, and the earth had lost its unique astronomical status, becoming one of a number of moving planets. Many of modern astronomy's principal achievements depend upon this transposition. A reform in the fundamental concepts of astronomy is therefore the first of the Copernican Revolution's meanings.

Astronomical reform is not, however, the Revolution's only meaning. Other radical alterations in man's understanding of nature rapidly followed the publication of Copernicus' *De Revolutionibus* in 1543. Many of these innovations, which culminated a century and a half later in the Newtonian conception of the universe, were unanticipated by-products of Copernicus' astronomical theory. Copernicus suggested the earth's motion in an effort to improve the techniques used in predicting the astronomical positions of celestial bodies. For other sciences his suggestion simply raised new problems, and until these were solved the astronomer's concept of the universe was incompatible with that of other scientists. During the seventeenth century, the reconciliation of these other sciences with Copernican astronomy was an important cause of the general intellectual ferment now known as the scientific revolution. Through the scientific revolution science won the great new role that it has since played in the development of Western society and Western thought.

Even its consequences for science do not exhaust the Revolution's meanings. Copernicus lived and worked during a period when rapid changes in political, economic, and intellectual life were preparing the bases of modern European and American civilization. His planetary theory and his associated conception of a sun-centered universe were instrumental in the transition from medieval to modern Western society, because they seemed to affect man's relation to the universe and to God. Initiated as a narrowly technical, highly mathematical revision of classical astronomy, the Copernican theory became one focus for the tremendous controversies in religion, in philosophy, and in social theory, which, during the two centuries following the discovery of America, set the tenor of the modern mind. Men who believed that their terrestrial home was only a planet circulating blindly about one of an infinity of stars evaluated their place in the cosmic scheme quite differently than had their predecessors who saw the earth as the unique and focal center of God's creation. The Copernican Revolution was therefore also part of a transition in Western man's sense of values.

This book is the story of the Copernican Revolution in all three of these not quite separable meanings — astronomical, scientific, and philosophical. The Revolution as an episode in the development of planetary astronomy will, of necessity, be our most developed theme. During the first two chapters, as we discover what the naked eye can see in the heavens and how stargazers first reacted to what they saw there, astronomy and astronomers will be very nearly our only concern. But once we have examined the main astronomical theories developed in the ancient world, our viewpoint will shift. In analyzing the strengths of the ancient astronomical tradition and in exploring the requisites for a radical break with that tradition, we shall gradually discover how difficult it is to restrict the scope of an established scientific concept to a single science or even to the sciences as a group. Therefore, in

Chapters 3 and 4 we shall be less concerned with astronomy itself than with the intellectual and, more briefly, the social and economic milieu within which astronomy was practiced. These chapters will deal primarily with the extra-astronomical implications — for science, for religion, and for daily life — of a time-honored astronomical conceptual scheme. They will show how a change in the conceptions of mathematical astronomy could have revolutionary consequences. Finally, in the last three chapters, when we turn to Copernicus' work, its reception, and its contribution to a new scientific conception of the universe, we shall deal with all these strands at once. Only the battle that established the concept of the planetary earth as a premise of Western thought can adequately represent the full meaning of the Copernican Revolution to the modern mind.

Because of its technical and historical outcome, the Copernican Revolution is among the most fascinating episodes in the entire history of science. But it has an additional significance which transcends its specific subject: it illustrates a process that today we badly need to understand. Contemporary Western civilization is more dependent, both for its everyday philosophy and for its bread and butter, upon scientific concepts than any past civilization has been. But the scientific theories that bulk so large in our daily lives are unlikely to prove final. The developed astronomical conception of a universe in which the stars, including our sun, are scattered here and there through an infinite space is less than four centuries old, and it is already out of date. Before that conception was developed by Copernicus and his successors, other notions about the structure of the universe were used to explain the phenomena that man observed in the heavens. These older astronomical theories differed radically from the ones we now hold, but most of them received in their day the same resolute credence that we now give our own. Furthermore, they were believed for the same reasons: they provided plausible answers to the questions that seemed important. Other sciences offer parallel examples of the transiency of treasured scientific beliefs. The basic concepts of astronomy have, in fact, been more stable than most.

The mutability of its fundamental concepts is not an argument for rejecting science. Each new scientific theory preserves a hard core of the knowledge provided by its predecessor and adds to it. Science progresses by replacing old theories with new. But an age as dominated by science as our own does need a perspective from which to examine the scientific beliefs which it takes so much for granted, and history provides one important source of such perspective. If we can discover the origins of some modern scientific concepts and the way in which they supplanted the concepts of an earlier age, we are more likely to evaluate intelligently their chances for survival. This book deals primarily with astronomical concepts, but they are much like those employed in many other sciences, and by scrutinizing their development we can learn something of scientific theories in general. For example: What is a scientific theory? On what should it be based to command our respect? What is its function, its use? What is its staying power? Historical analysis may not answer questions like these, but it can illuminate them and give them meaning.

Because the Copernican theory is in many respects a typical scientific theory, its history can illustrate some of the processes by which scientific concepts evolve and replace their predecessors. In its extrascientific consequences, however, the Copernican theory is not typical: few scientific theories have played so large a role in non-scientific thought. But neither is it unique. In the nineteenth century, Darwin's theory of evolution raised similar extrascientific questions. In our own century, Einstein's relativity theories and Freud's psycho-analytic theories provide centers for controversies from which may emerge further radical reorientations of Western thought. Freud himself emphasized the parallel effects of Copernicus' discovery that the earth was merely a planet and his own discovery that the unconscious controlled much of human behavior. Whether we have learned their theories or not, we are the intellectual heirs of men like Copernicus and Darwin. Our fundamental thought processes have been reshaped by them, just as the thought of our children or grandchildren will have been reshaped by the work of Einstein and Freud. We need more than an understanding of the internal development of science. We must also understand how a scientist's solution of an apparently petty, highly technical problem can on occasion fundamentally alter men's attitudes toward basic problems of everyday life.

The Heavens in Primitive Cosmologies

Much of this book will deal with the impact of astronomical observations and theories upon ancient and early modern cosmological thought, that is, upon a set of man's conceptions about the structure of the universe. Today we take it for granted that astronomy should affect cosmology. If we want to know the shape of the universe, the earth's position in it, or the relation of the earth to the sun and the sun to the stars, we ask the astronomer or perhaps the physicist. They have made detailed quantitative observations of the heavens and the earth; their knowledge of the universe is guaranteed by the accuracy with which they predict its behavior. Our everyday conception of the universe, our popular cosmology, is one product of their painstaking researches. But this close association of astronomy and cosmology is both temporally and geographically local. Every civilization and culture of which we have records has had an answer for the question, "What is the structure of the universe?" But only the Western civilizations which descend from Hellenic Greece have paid much attention to the appearance of the heavens in arriving at that answer. The drive to construct cosmologies is far older and more primitive than the urge to make systematic observations of the heavens. Furthermore, the primitive form of the cosmological drive is particularly informative because it highlights features obscured in the more technical and abstract cosmologies that are familiar today.

Though primitive conceptions of the universe display considerable substantive variation, all are shaped primarily by terrestrial events, the events that impinge most immediately upon the designers of the systems. In such cosmologies the heavens are merely sketched in to provide an enclosure for the earth, and they are peopled with and moved by mythical figures whose rank in the spiritual hierarchy usually increases with their distance from the immediate terrestrial environment. For example, in one principal form of Egyptian cosmology the earth was pictured as an elongated platter. The platter's long dimension paralleled the Nile; its flat bottom was the alluvial basin to which ancient Egyptian civilization was restricted; and its curved and rippled rim was the mountains bounding the terrestrial world. Above the platter-earth was air, itself a god, supporting an inverted platter-dome which was the skies. The terrestrial platter in its turn was supported by water, another god, and the water rested upon a third platter which bounded the universe symmetrically from below.

Clearly several of the main structural features of this universe were suggested by the world that the Egyptian knew: he did live in an elongated platter bounded by water in the only direction in which he had explored it; the sky, viewed on a clear day or night, did and does look dome-shaped; a symmetric lower boundary for the universe was the obvious choice in the absence of relevant observations. Astronomical appearances were not ignored, but they were treated with less precision and more myth. The sun was Ra, the principal Egyptian god, supplied with two boats, one for his daily journey through the air and a second for his nocturnal trip through the water. The stars were painted or studded in the vault of the heaven; they moved as minor gods; and in some versions of the cosmology they were reborn each night. Sometimes more detailed observations of the heavens entered, as when the circumpolar stars (stars that never dip below the horizon) were recognized as "those that know no weariness" or "those that know no destruction." From such observations the northern heavens were identified as a region where there could be no death, the region of the eternally blessed afterlife. But such traces of celestial observation were rare.

Fragments of cosmologies similar to the Egyptian can be found in all those ancient civilizations, like India and Babylonia, of which we have records. Other crude cosmologies characterize the contemporary primitive societies investigated by the modern anthropologist. Apparently all such sketches of the structure of the universe fulfill a basic psychological need: they provide a stage for man's daily activities and the activities of his gods. By explaining the physical relation between man's habitat and the rest of nature, they integrate the universe for man and make him feel at home in it. Man does not exist for long without inventing a cosmology, because a cosmology can provide him with a world-view which permeates and gives meaning to his every action, practical and spiritual.

Though the psychological needs satisfied by a cosmology seem relatively uniform, the cosmologies capable of fulfilling these needs have varied tremendously from one society or civilization to another. None of the primitive cosmologies referred to above will now satisfy our demand for a world-view, because we are members of a civilization that has set additional standards which a cosmology must meet in order to be believed. We will not, for example, credit a cosmology that employs gods to explain the everyday behavior of the physical world; in recent centuries, at least, we have insisted upon more nearly mechanical explanations. Even more important, we now demand that a satisfactory cosmology account for many of the observed details of nature's behavior. Primitive cosmologies are only schematic sketches against which the play of nature takes place; very little of the play is incorporated into the cosmology. The sun god, Ra, travels in his boat across the heavens each day, but there is nothing in Egyptian cosmology to explain either the regular recurrence of his journey or the seasonal variation of his boat's route. Only in our own Western civilization has the explanation of such details been considered a function of cosmology. No other civilization, ancient or modern, has made a similar demand.

The requirement that a cosmology supply *both* a psychologically satisfying world-view *and* an explanation of observed phenomena like the daily change in the position of sunrise has vastly increased the power of cosmologic thought. It has channeled the universal compulsion for at-homeness in the universe into an unprecedented drive for the discovery of scientific explanations. Many of the most characteristic achievements of Western civilization depend upon this combination of demands imposed upon cosmologic thought. But the combination has not always been a congenial one. It has forced modern man to delegate the construction of cosmologies to specialists, primarily to astronomers, who know the multitude of detailed observations that modern cosmologies must satisfy to be believed. And since observation is a two-edged sword which may either confirm or conflict with a cosmology, the consequences of this delegation can be devastating. The astronomer may on occasions destroy, for reasons lying entirely within his specialty, a world-view that had previously made the universe meaningful for the members of a whole civilization, specialist and nonspecialist alike.

Something very much like this happened during the Copernican Revolution. To understand it we must therefore become something of specialists ourselves. In particular, we must get to know the principal observations, all of them accessible to the naked eye, upon which depend the two main scientific cosmologies of the West, the Ptolemaic and the Copernican. No single panoramic view of the heavens will suffice. Seen on a clear night, the skies speak first to the poetic, not to the scientific, imagination. No one who views the night sky can challenge Shakespeare's vision of the stars as "night's candles" or Milton's image of the Milky Way as "a broad and ample road, whose dust is gold, and pavement stars." But these descriptions are the ones embodied in primitive cosmologies. They provide no evidence relevant to the astronomer's questions: How far away is the Milky Way, the sun, the planet Jupiter? How do these points of light move? Is the material of the moon like the earth's, or is it like the sun's, or like a star's? Questions like these demand systematic, detailed, and quantitative observations accumulated over a long period of time.

This chapter deals, then, with observations of the sun and stars and with the role of these observations in establishing the first scientific cosmologies of ancient Greece. The next chapter completes the roster of naked-eye celestial observations by describing the planets, the celestial bodies which posed the technical problem that led to the Copernican Revolution.

The Apparent Motion of the Sun

Before the end of the second millennium B.C. (perhaps very much before), the Babylonians and the Egyptians had begun systematic observations of the motion of the sun. For this purpose they developed a primitive sundial consisting of a measured stick, the gnomon, projecting vertically from a smooth flat section of ground. Since the apparent position of the sun, the tip of the gnomon, and the tip of its shadow lie along a straight line at each instant of a clear day, measurements of the length and direction of the shadow completely determine the direction of the sun. When the shadow is short, the sun is high in the sky; when the shadow points, say, to the east, the sun must lie in the west. Repeated observations of the gnomon's shadow can therefore systematize and quantify a vast amount of common but vague knowledge about the daily and annual variation of the sun's position. In antiquity such observations harnessed the sun as a time reckoner and calendar keeper, applications that provided one important motive for continuing and refining the observational techniques.

Both the length and the direction of a gnomon's shadow vary slowly and continuously during the course of any one day. The shadow is longest at sunrise and sunset, at which times it points in roughly opposite directions. During the daylight hours the shadow moves gradually through a symmetric fan-shaped figure which, in most of the locations accessible to ancient observers, is much like one of those shown in Figure 1. As the diagram indicates, the shape of the fan is different on different days, but it has one very significant fixed feature. At the instant of each day when the gnomon's shadow is shortest, it always points in the same direction. This simple regularity provides two fundamental frames of reference for all further astronomical measurements. The permanent direction assumed by the shortest shadow each day defines due north, from which the other compass points follow; the instant at which the shadow becomes shortest defines a reference point in time, local noon; and the interval between two successive local noons defines a fundamental time unit, the apparent solar day. During the first millennium B.C., the Babylonians, Egyptians, Greeks, and Romans used primitive terrestrial timekeepers, particularly water clocks, in order to subdivide the solar day into smaller intervals from which our modern units of time — hour, minute, and second — descend.°

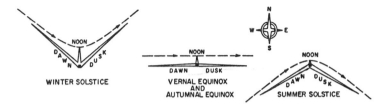

Figure 1. The daily motion of the gnomon's shadow at various seasons in middle-northern latitudes. At sunrise and sunset the shadow stretches momentarily to infinite distance where its end "joins" the broken line in the diagram. Between sunrise and sunset the end of the shadow moves slowly along the broken line; at noon the shadow always points due north.

The compass points and the time units defined by the sun's daily motion provide a basis for describing the changes in that motion from day to day. Sunrise always occurs somewhere in the east and sunset in the west, but the position of sunrise, the length of the gnomon's noon shadow, and the number of daylight hours vary from day to day with the changing seasons (Figure 2). The winter solstice is the day (December 22 on the modern calendar) when the sun rises and sets farthest to the south of the due east and west points on the horizon. On this day there are fewer hours of daylight and the gnomon's noon shadow is longer than on any other. After the winter solstice the points

° For astronomical purposes the stars provide a more convenient time reckoner than the sun. But, on a time scale determined by the stars, the length of the apparent solar day varies by almost a minute at different seasons of the year. Though ancient astronomers were aware of this slight but significant irregularity of apparent solar time, we shall ignore it here. The cause of this variation and its effect upon the definition of a time scale are discussed in Section 1 of the Technical Appendix.

at which the sun rises and sets gradually move north together along the horizon, and the noon shadows grow shorter. On the vernal equinox (March 21) the sun rises and sets most nearly due east and west; nights and days are then of equal length. As more days pass, the sunrise and sunset points continue to move northward and the number of daylight hours increases until the summer solstice (June 22), when the sun rises and sets farthest to the north. This is the time when daylight lasts longest and when the gnomon's noon shadow is shortest. After the summer solstice, the sunrise point again moves south, and the nights grow longer. At the autumnal equinox (September 23) the sun once again rises and sets almost due east and west; then it continues south until the winter solstice recurs.

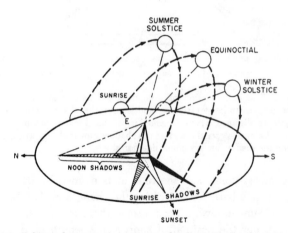

Figure 2. Relation between the position of sunrise, the sun's noon elevation, and the seasonal variation of the gnomon's shadow.

As the modern names of the solstices and equinoxes indicate, the motion of sunrise back and forth along the horizon corresponds to the cycle of the seasons. Most ancient peoples therefore believed that the sun controlled the seasons. They simultaneously venerated the sun as a god and observed it as a calendar keeper, a practical indicator of the passage of the seasons upon which their agricultural activities depended. Prehistoric remains, like the mysterious structure of giant stones at Stonehenge, England, testify to the antiquity and the strength of this double interest in the sun. Stonehenge was an important temple laboriously constructed from huge stones, some almost thirty tons in weight, by the people of an early Stone Age civilization. It was almost certainly also a crude sort of observatory. The stones were so arranged that an observer at the center of the array saw the sun rise over a specially placed stone, called the "Friar's Heel," on the ancient midsummer day, the summer solstice.

The length of the cycle of the seasons — the interval between one vernal equinox and the next — defines the basic calendar unit, the year, just as the sun's daily motion defines the day. But the year is a far more difficult unit to measure than the day, and the demand for useful long-term calendars has therefore presented astronomers with a continuing problem whose prominence during the sixteenth century played a direct role in the Copernican Revolution. The earliest solar calendars of antiquity were based upon a year of 360 days, a neat round number that nicely fitted the sexagesimal number system of the Babylonians. But the cycle of the seasons occupies more than 360 days, so that the "New Year's Day" of these early solar calendars gradually crept around the cycle of the seasons from winter, to fall, to summer, to spring. The calendar was scarcely useful over long periods of time, because important seasonal events, like the flooding of the Nile in Egypt, occurred at later and later dates in successive years. To keep the solar calendar in step with the seasons, the Egyptians therefore added five extra days, a holiday season, to their original year.

There is, however, no integral number of days in the cycle of the seasons. The year of 365 days is also too short, and after 40 years the Egyptian calendar was ten days out of step with the seasons. Therefore, when Julius Caesar reformed the calendar with technical assistance from Egyptian astronomers, he based his new calendar upon a year 365¼ days in length; three years of 365 days were followed by one of 366. This calendar, the Julian, was used throughout Europe from its introduction in 45 B.C. until after the death of Copernicus. But the seasonal year is actually 11 minutes and 14 seconds shorter than 365¼ days, so that by Copernicus' lifetime the date of the vernal equinox had moved backward from March 21 to March 11. The resulting demand for calendar reform (see Chapters 4 and 5) provided one important motive for the reform of astronomy itself, and the reform that gave the Western world its modern calendar followed the publication of the *De Revolutionibus* by only thirty-nine years. In the new calendar, imposed upon large areas of Christian Europe by Pope Gregory XIII in 1582, leap year is suppressed three times in every four hundred years. The year 1600 was a leap year and the year 2000 will be, but 1700, 1800, and 1900, all leap years in the Julian calendar, had just 365 days in the Gregorian, and 2100 will again be a normal year of 365 days.

All the observations discussed above show the sun approximately as it would appear to an astronomer in middle-northern latitudes, an area that includes Greece, Mesopotamia, and northern Egypt, the regions in which almost all ancient observations were made. But within this area there is a considerable quantitative variation in certain aspects of the sun's behavior, and in the southernmost parts of Egypt there is a qualitative change as well. Knowledge of these changes also played a part in the construction of ancient astronomical theories. No variations are observed as an observer moves east or west. But toward the south the noon shadow of the gnomon is shorter and the noon sun higher in the sky than they would be on the same day in the north. Similarly, though the length of the whole day remains constant, the difference between the lengths of daytime and nighttime is smaller in the southern portion of middle-northern latitudes. Also, in this region the sun does not swing quite so far north and south along the horizon

Figure 3. The daily motion of the gnomon's shadow at various seasons in the northern torrid zone.

during the course of the year. None of these variations alters the qualitative descriptions supplied above. But, if an observer has moved far into southern Egypt during the summer, he will see the noon shadow of the gnomon grow shorter day by day until at last it vanishes entirely and then reappears pointing to the south. In the southernmost parts of Egypt the annual behavior of the gnomon's shadow is that shown in Figure 3. Journeys still farther south or much farther north will produce other anomalies in the observed motions of the sun. But these were not observed in antiquity. We shall not discuss them until we deal with the astronomical theories that made it possible to predict them even before they were observed (pp. 33 ff).

The Stars

The motions of the stars are much simpler and more regular than the sun's. Their regularity is not, however, so easily recognized because systematic examination of the night sky requires the ability to select individual stars for repeated study wherever in the heavens they appear. In the modern world this ability, which can be acquired only by long practice, is quite rare. Few people now spend much time out of doors at night, and, when they do, their view of the heavens is frequently obscured by tall buildings and street lighting. Besides, observation of the heavens no longer has a direct role in the life of the average man. But in antiquity the stars were an immediate part of the normal man's environment, and celestial bodies served a universal function as time reckoners and calendar keepers. Under these circum-

Figure 4. The constellation Ursa Major in the northern skies. Notice the familiar Big Dipper whose handle forms the bear's tail. The North Star is the prominent star directly over the bear's right ear in the picture. It lies almost on a line joining the last two stars in the Dipper's bowl.

stances the ability to identify stars at a glance was relatively common. Long before the beginning of recorded history men whose jobs gave them a continuing view of the night sky had mentally arranged the stars into constellations, groups of neighboring stars that could be seen and recognized as a fixed pattern. To find an individual star amidst the profusion of the heavens, an observer would first locate the familiar star pattern within which it occurred and then pick the individual star from the pattern.

Many of the constellations used by modern astronomers are named after mythological figures of antiquity. Some can be traced to Babylonian cuneiform tablets, a few as old as 3000 B.C. Though modern astronomy has modified their definitions, the major constellations are among our oldest traceable inheritances. How these groups were first picked out is, however, still uncertain. Few people can "see" a bear in the stars of the constellation Ursa Major (Figure 4); other constellations present similar problems in visualization; the stars may therefore originally have been grouped for convenience and named arbitrarily. But, if so, they were very strangely grouped. The ancient constellations have very irregular boundaries, and they occupy areas of quite different size in the sky. They are not convenient choices, which is one reason why modern astronomers have altered their boundaries. Probably the ancient shepherd or navigator, staring at the heavens hour after hour, really did "see" his familiar mythological characters traced in the stars, just as we sometimes "see" faces in clouds or the outlines of trees. The experiments of modern Gestalt psychology demonstrate a universal need to discover familiar patterns in apparently random groupings, a need that underlies the well-known "ink-blot" or Rohrschach tests. If we knew more about their historical origin, the constellations might provide useful information about the mental characteristics of the prehistoric societies that first traced them.

Learning the constellations is like gaining familiarity with a map and has the same purpose: the constellations make it easier to find one's way around the sky. Knowing the constellations, a man can readily find a comet reported to be "in Cygnus" (the Swan); he would almost certainly miss the comet if he knew only that it was "in the sky." The map provided by the constellations is, however, an unusual one because the constellations are always in motion. Since they all move together, preserving their patterns and their relative positions, the motion does not destroy their usefulness. A star in Cygnus will always be in Cygnus, and Cygnus will always be the same distance from Ursa Major.° But neither Cygnus nor Ursa Major remains for long at the same position in the sky. They behave like cities on a map pasted to a rotating phonograph record.

Both the fixed relative positions and the motions of the stars are illustrated in Figure 5, which shows the location and orientation of the Big Dipper (part of Ursa Major) in the northern sky at three times during a single night. The pattern of seven stars in the Dipper is the same at each viewing. So is the relation of the Dipper to the North Star, which always lies 29° to the open side of the Dipper's bowl on a straight line through the last two stars in the bowl. Other diagrams would show similar permanent geometric relations among the other stars in the heavens.

Figure 5 displays another important characteristic of the stellar motions. As the constellations and the stars composing them swing through the skies together, the North Star remains very nearly stationary. Careful observation shows that it is not, in fact, quite stationary during any night, but there is another point in the heavens, now less than 1° away from the North Star, which has precisely the properties attributed to that star in Figure 5. This point is known as the north celestial pole. An observer at a given location in northern latitudes can always find it, hour after hour and night after night, at the same fixed distance above the due-north point on his horizon. A straight stick clamped so that it points toward the pole will continue to point toward the pole as the stars move. Simultaneously, however, the celestial pole behaves as a star. That is, the pole retains its geometric relations to the stars over long periods of time.† Since the pole is a fixed

° "Distance" here means "angular distance," that is, the number of degrees between two lines pointing from the observer's eye to the two celestial objects whose separation is to be measured. This is the only sort of distance that astronomers can measure directly, that is, without making calculations based upon some theory about the structure of the universe.

† Observations made many years apart show that the pole's position among the stars is very slowly changing (about 1° in 180 years). We shall neglect this slow motion, which is part of an effect known as the precession of the equinoxes, until Section 2 of the Technical Appendix. Though the ancients were aware of it by the end of the second century B.C., precession played a secondary role in the construction of their astronomical theories, and it does not alter the short-term observations described above. There has always been a north celestial pole at the same distance above the due-north point on the horizon, but the same stars have not always been near it.

point for each terrestrial observer and since the stars do not change their distance from this point as they move, every star seems to travel along the arc of a circle whose center is the celestial pole. Figure 5 shows a part of this circular motion for the stars in the Dipper.

The concentric circles traced by the circumpolar motions of the stars are known as their diurnal or daily circles, and the stars revolve in these circles at a rate just over 15° per hour. No star completes a full circle between sunset and sunrise, but a man observing the northern skies during a single clear night can follow stars near the pole through approximately a semicircle, and on the next night he can find them again moving along the same circles at the same rate. Furthermore, he will find them at just the positions they would have reached if they had continued their steady revolutions throughout the intervening day. Since antiquity, most observers equipped to recognize these regularities have naturally assumed that the stars exist and move during the day as during the night, but that during the day the strong light of the sun makes them invisible to the naked eye. On this interpretation the stars swing steadily through full circles, each star completing a circle once every 23 hours 56 minutes. A star that is directly below the pole at 9:00 o'clock on the evening of October 23 will return to the same position at 8:56 on the evening of October 24

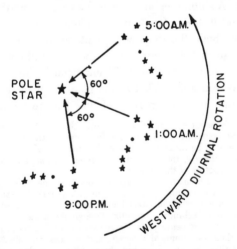

Figure 5. Successive positions of the Big Dipper at four-hour intervals on a night in late October.

and at 8:52 on October 25. By the end of the year it will be reaching its position below the pole before sunset and will therefore not be visible in that position at all.

In middle-northern latitudes the celestial pole is approximately 45° above the northernmost point on the horizon. (The elevation of the pole is precisely equal to the observer's angle of latitude — that is one way latitude is measured.) Therefore stars that lie within 45° of the pole, or whatever the elevation at the observer's location may be, can never fall below the horizon and must be visible at any hour of a clear night. These are the circumpolar stars, "those that know no destruction," in the words of the ancient Egyptian cosmologists. They are also the only stars whose motion is easily recognized as circular.

Stars farther from the poles also travel along diurnal circles, but part of each circle is hidden below the horizon (Figure 6). Therefore such stars can sometimes be seen rising or setting, appearing above or disappearing below the horizon; they are not always visible throughout the night. The farther from the pole such a star is, the less of its diurnal circle is above the horizon and the more difficult it is to recognize the visible portion of its path as part of a circle. For example, a star that rises due east is visible on only half of its diurnal circle. It travels very nearly the same path that the sun takes near one of the equinoxes, rising along a slant line up and to the south (Figure 7a), reaching its maximum height at a point over the right shoulder of an observer looking east, and finally setting due west along a line slanting downward and to the north. Stars still farther from the pole appear only briefly over the southern horizon. Near the due-south point they set very soon after they rise, and they never get very far above the horizon (Figure 7b). Since during almost half the year they rise and set during daylight, there are many nights when they do not appear at all.

These qualitative features of the night sky are common to the entire area within which ancient astronomical observations were made, but the description has glossed over significant quantitative differences. As an observer travels south, the elevation of the pole above the northern horizon decreases approximately 1° for every 69 miles of southward motion. The stars continue to move in diurnal circles about the pole, but since the pole is closer to the horizon, some stars that were circumpolar in the north are seen rising and setting by an observer farther south. Stars that rise and set due east and west continue to appear and disappear at the same points on the horizon, but toward the south they appear to move along a line more nearly perpendicular to the horizon, and they reach their maximum elevation more nearly

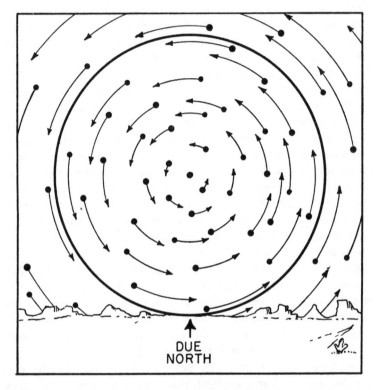

Figure 6. A set of the short circular arcs described by typical stars in the northern sky during a two-hour period. The heavy circle tangent to the horizon separates the circumpolar stars from those that rise and set.

Star trails like these can actually be recorded by pointing a fixed camera *at the celestial pole* and leaving the shutter open as the heavens turn. Each additional hour's exposure adds 15° to the length of every track. Notice, however, that the elevated camera angle introduces a deceptive distortion. If the pole is 45° above the horizon (a typical elevation in middle-northern latitudes), then a star that appears at the very top of the heavy circle is actually directly above the observer's head. Recognizing the distortion due to camera angle makes it possible to relate the star trails in this diagram to those shown more schematically in Figures 7a and b.

over the observer's head. The appearance of the southern sky changes more strikingly. As the pole declines toward the northern horizon, stars in the southern sky, because they remain at the same angular distance from the pole, rise to greater heights over the southern horizon. A star that barely rises above the horizon when seen from the north will rise higher and be seen for longer when observed from farther south. A southern observer will still see stars that barely peek above the southernmost point on his horizon, but these will be stars that the northern stargazer never sees at all. As an observer moves south, he sees fewer and fewer circumpolar stars — stars that are visible through-

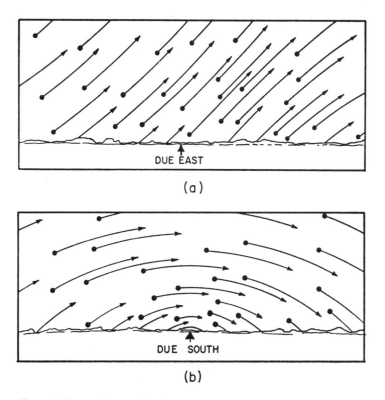

(a)

(b)

Figure 7. Star trails over (a) the eastern and (b) the southern horizon. Like Figure 6, these diagrams show the motion of typical stars over a 90° section of the horizon during a two-hour period. In these diagrams, however, the "camera" is directed to the horizon, so that only the first 40° above the horizon is shown.

out the night. But in the south he will, at some time or other, observe stars that an observer in the north can never see.

The Sun as a Moving Star

Because the stars and the celestial pole retain the same relative positions hour after hour and night after night, they can be permanently located upon a map of the heavens, a star map. One form of star map is shown in Figure 8; others will be found in any atlas or book on astronomy. The map of Figure 8 contains all the brighter stars that can ever be seen by an observer in middle-northern latitudes, but not all the stars on the map can be seen at once because they are not all above the horizon simultaneously. At any instant of the night approximately two-fifths of the stars on the map lie below the horizon.

The particular stars that are visible and the portion of sky in which they appear depend upon the date and hour of the observation. For example, the solid black line on the map broken by the four cardinal points of the compass, N, E, S, W, encloses the portion of the sky that is visible to an observer in middle-northern latitudes at 9:00 o'clock on the evening of October 23. It therefore represents his horizon. If the observer holds the map over his head with the bottom toward the north, the four compass points will be approximately aligned with the corresponding points on his physical horizon. The map then indicates that at this time of night and year the Big Dipper appears just over the northern horizon and that, for example, the constellation Cassiopeia lies at a position near the center of the horizon-window, corresponding to a position nearly overhead in the sky. Since the stars return to their positions in just 4 minutes less than 24 hours, the same orientation of the map must indicate the position of the stars at 8:56 on the evening of October 24, at 8:52 on October 25, at 8:32 on October 30, and so on.

Now imagine that the solid black horizon line which encloses the observer's field of view is held in its present position on the page while the entire disk of the map is rotated slowly behind it in a counterclockwise direction about the central pole. Rotating the disk 15° brings into the horizon-window just those stars that are visible at 10:00 o'clock on the evening of October 23, or at 9:56 on the evening of October 24, and so on. A rotation of 45° moves the stars visible at midnight on October 23 inside of the horizon line. The positions of all bright

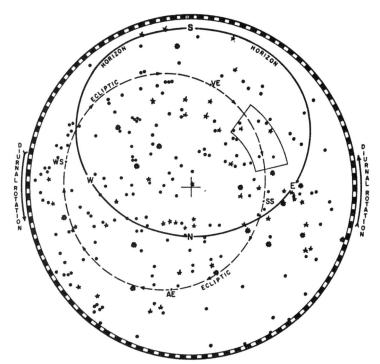

Figure 8. A circumpolar star map containing all the major stars ever visible to an observer at approximately 45° northern latitude. The cross at the geometric center of the map indicates the position of the celestial pole.

If the map is held horizontally overhead with its face toward the ground and with the bottom of the page pointing north, it will show the orientation of the stars as they appear to an observer in middle-northern latitudes at 9:00 o'clock on the evening of October 23. The stars within the solid line bounding the horizon-window are the ones that the observer can see; those outside the line are below the horizon on this day at this hour. Stars that lie within the horizon-window near the point *N* on the map will be seen just over the due-north point on the physical horizon (notice the Dipper); those near the east point, *E*, will be just rising in the east; and so on. To find the position of stars at a later hour on October 23, the horizon-window should be imagined stationary and the circular map should be rotated behind it, counterclockwise about the pole, 15° for each hour after 9:00 P.M. This motion leaves the pole stationary but carries stars up over the eastern horizon and down behind the western one. To find the positions of stars at 9:00 P.M. on a later day the map should be rotated clockwise behind the stationary horizon-window, 1° for each day after October 23. Combining these two procedures makes it possible to find the positions of stars at any hour of any night of the year.

The broken line that encircles the pole in the diagram is the ecliptic, the sun's apparent path through the stars (see p. 23). The box that encloses a portion of the ecliptic in the upper right-hand quadrant of the map contains the region of the sky shown in expanded form in Figures 9 and 15.

stars at any hour of any night can be found in this way. A movable
star map equipped with a fixed horizon-window, like that in Figure 8,
is frequently known as a "star finder."

Star maps have other applications, however, besides locating bodies
that, like the stars, remain in constant relative positions. They can
also be used to describe the behavior of celestial bodies that, like the
moon, comets, or planets, slowly change their positions among the
stars. For example, as the ancients knew, the sun's motion takes a
particularly simple form as soon as it is related to the stars. Since the
stars appear shortly after sunset, an observer who knows how to follow

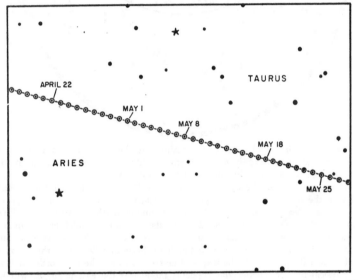

Figure 9. Motion of the sun through the constellations Aries and Taurus. The
circles represent the sun's position among the stars at sunset on successive evenings
from the middle of April to late May.

their motion can record the time and horizon position of the sunset,
measure the time between sunset and the first appearance of the stars,
and then locate the sun on a star map by rotating the map backward
to determine which stars were at the appropriate horizon position when
the sun set. An observer who plots the position of the sun on a star
map for several consecutive evenings will find it in almost the same
position each time. Figure 9 shows the position of the sun on a star

map on successive evenings for one month. It is not in the same position on the map for two successive observations, but it has not moved far. Each evening finds it about 1° from its position the previous evening, and 1° is a relatively small distance, about twice the angular diameter of the sun.

These observations suggest that both the daily motion of the sun and its slower shift north and south along the horizon may conveniently be analyzed by regarding the sun as a body that moves slowly among the stars from day to day. If, for some particular day, the position of the sun among the stars is specified, then, on that day, the sun's motion will be almost exactly the diurnal motion of a star in the corresponding position on the map. Both will move like points on the rotating map, rising in the east along a line slanting upward and to the south and later setting in the west. One month later the sun will again have the diurnal motion of a star, but now it will move very nearly like a star 30° away from the position of the star whose motion it copied a month earlier. During the intervening month the sun has moved slowly and steadily between these two positions, 30° apart on the map. Each day its motion has been almost that of a star, part of a circle about the pole of the heavens, but it has not behaved like quite the same star on two successive days.

If the sun's position is plotted on a star map day after day and the points marking its successive positions in the evening are connected together, a smooth curve is produced which rejoins itself at the end of a year. This is the curve, called the ecliptic, that is indicated by the broken line on the star map of Figure 8. The sun is always to be found somewhere on this line. As the ecliptic is carried rapidly through the heavens by the common diurnal motion of the stars, the sun is carried along with it, rising and setting like a star located at a point somewhere on the line. But simultaneously the sun is moving slowly around the ecliptic, occupying a slightly different position each day, hour, or minute. Thus the complex helical motion of the sun can be analyzed as the result of two much simpler motions. The total apparent motion of the sun is composed of its diurnal motion (the westward circle due to the counterclockwise motion of the whole map) and a simultaneous slow eastward motion (clockwise about the pole on the map) along the ecliptic.

Analyzed in this way, the sun's motion shows close parallels to the motion of a toll collector on a merry-go-round. The collector is carried around rapidly by the revolutions of the platform. But as he walks slowly from horse to horse collecting tolls his motion is not quite the same as that of the riders. If he walks in a direction opposite to that of the platform's spin, his motion over the ground will be slightly slower than that of the platform, and the riders will complete one circle somewhat more rapidly than he. If his toll collections carry him toward and away from the center of the platform, his total motion with respect to the ground will not be circular at all, but a complex curve which does not rejoin itself at the end of a single revolution. Though it is theoretically possible to specify precisely the path along which the toll collector moves over the stationary ground, it is far simpler to divide his total motion into its two component parts: a steady rapid rotation with the platform and a slower less regular motion with respect to the platform. Since antiquity astronomers have used a similar division in analyzing the apparent motion of the sun. Each day the sun moves rapidly westward *with the stars* (its so-called diurnal motion); simultaneously the sun moves slowly eastward along the ecliptic *through the stars* or *with respect to the stars* (its annual motion).

With the sun's total motion divided into two components, its behavior can be described simply and precisely merely by labeling neighboring points on the ecliptic with the day and hour at which the sun reaches each of them. The series of labeled points specifies the annual component of the sun's motion; the remaining diurnal component is specified by the daily rotation of the map as a whole. For example, since the ecliptic appears in Figure 8 as a somewhat distorted and considerably off-center circle, there must be one point, SS, on the ecliptic that is nearer the central pole than any other. No other point on the ecliptic rises and sets as far to the north as SS, and no other joint stays within the horizon-window for as long during the map's rotation. Therefore SS is the summer solstice, and the sun's center must pass through it around June 22. Similarly the points *AE* and *VE* in Figure 8 are the equinoctial points, the two points on the ecliptic that rise and set due east and west and that remain inside the horizon-window for exactly one-half of each map rotation. The center of the sun must pass through them on September 23 and March 21 respectively, just as it must pass through WS, the point on the ecliptic farthest from the pole, on or near December 22. The solstices and equinoxes, which first appeared as days of the year, have now received a more precise and astronomically more useful definition. They are points on a star map or in the sky. Together with the corresponding dates (or instants, since the sun's center passes instantaneously through each point), these labeled positions on the ecliptic specify the direction and approximate rate of the sun's annual motion. Given these labels and others like them, a man who knows how to simulate the diurnal motion by rotating a star map can determine the hours and positions of sunrise and sunset and the maximum height of the sun on every day of the year.

The solstices and the equinoxes are not the only positions on the ecliptic to receive standard labels. Drawn on a star map, the ecliptic passes through a group of particularly prominent constellations, known as the signs of the zodiac. By a convention dating from remote antiquity these signs divide the ecliptic into twelve segments of equal length. To say that the sun is "in" a particular constellation is to specify approximately its position on the ecliptic, which, in turn, specifies the season of the year. The annual journey of the sun through the twelve signs seems to control the cycle of the seasons, an observation that is one root of the science or pseudo science of astrology with which we shall deal further in Chaper 3.

The Birth of Scientific Cosmology — The Two-Sphere Universe

The observations described in the last three sections are an important part of the data used by ancient astronomers in analyzing the structure of the universe. Yet, in themselves, these observations provide no direct structural information. They tell nothing about the composition of the heavenly bodies or their distance; they give no explicit information about the size, position, or shape of the earth. Though the method of reporting the observations has disguised the fact, they do not even indicate that the celestial bodies really move. An observer can only be sure that the angular distance between a celestial body and the horizon changes continually. The change might as easily be caused by a motion of the horizon as by a motion of the heavenly body. Terms like sunset, sunrise, and diurnal motion of a star do not, strictly speaking, belong in a record of observation at all. They are parts of an interpretation of the data, and though this interpretation is so natural that it can scarcely be kept out of the vocabulary with which the observations are discussed, it does go beyond the content of the observations themselves. Two astronomers can agree perfectly about the results of observation and yet disagree sharply about questions like the reality of the motion of the stars.

Observations like those discussed above are therefore only clues to a puzzle for which the theories invented by astronomers are tentative solutions. The clues are in some sense objective, given by nature; the numerical result of this sort of observation depends very little upon the imagination or personality of the observer (though the way in which the data are arranged may). But the theories or conceptual schemes derived from these observations do depend upon the imagination of scientists. They are subjective through and through. Therefore, observations like those discussed in the preceding sections could be collected and put in systematic form by men whose beliefs about the structure of the universe resembled those of the ancient Egyptians. The observations in themselves have no *direct* cosmological consequences; they need not be, and for many millennia were not, taken very seriously in the construction of cosmologies. The tradition that detailed astronomical observations supply the principal clues for cosmological thought is, in its essentials, native to Western civilization. It seems to be one of the most significant and characteristic novelties that we inherit from the civilization of ancient Greece.

A concern to explain observations of the stars and planets is apparent in our oldest fragmentary records of Greek cosmological thought. Early in the sixth century B.C., Anaximander of Miletus taught:

> The stars are compressed portions of air, in the shape of [rotating] wheels filled with fire, and they emit flames at some point from small openings. . . .
>
> The sun is a circle twenty-eight times the size of the earth; it is like a chariot-wheel, the rim of which is hollow and full of fire, and lets the fire shine out at a certain point in it through an opening like the nozzle of a pair of bellows. . . .
>
> The eclipses of the sun occur through the orifice by which the fire finds vent being shut up.
>
> The moon is a circle nineteen times as large as the earth; it is like a chariot-wheel, the rim of which is hollow and full of fire, like the circle of the sun, and it is placed obliquely, as that of the sun also is; it has one vent like the nozzle of a pair of bellows; its eclipses depend on the turnings of the wheel.[1]

Astronomically these conceptions are far in advance of the Egyptians'. The gods have vanished in favor of mechanisms familiar on the earth. The size and position of the stars and planets are discussed. Though the answers given seem extremely rudimentary, the problems had to be raised before they could receive mature and considered solutions. In the fragment quoted the diurnal circles of the stars and the sun are handled with some success by treating the celestial bodies as orifices on the rims of rotating wheels. The mechanisms for eclipses and for the annual wandering of the sun (the latter accounted for by the oblique position of the sun's circle) are less successful, but they are at least begun. Astronomy has started to play a major role in cosmological thought.

Not all the Greek philosophers and astronomers agreed with Anaximander. Some of his contemporaries and successors advanced other theories, but they advanced them for the same problems and they employed the same techniques in arriving at solutions. For us it is the problems and techniques that are important. The competing theories need not be traced; moreover, they cannot be traced completely, for the historical records are too incomplete to permit more than conjecture about the evolution of the earliest Greek conceptions of the universe. Only in the fourth century B.C. do the records become approximately reliable, and by that time, as the result of a long evolutionary process, a large measure of agreement about cosmological essentials had been reached. For most Greek astronomers and philosophers, from the fourth century on, the earth was a tiny sphere suspended stationary at the geometric center of a much larger rotating sphere which carried the stars. The sun moved in the vast space between the earth and the sphere of the stars. Outside of the outer sphere there was nothing at all — no space, no matter, nothing. This was not, in antiquity, the only theory of the universe, but it is the one that gained most adherents, and it is a developed version of this theory that the medieval and modern world inherited from the ancients.

This is what I shall henceforth call the "two-sphere universe," consisting of an interior sphere for man and an exterior sphere for the stars. The phrase is, of course, an anachronism. As we shall see in the next chapter, all those philosophers and astronomers who believed in the terrestrial and celestial spheres also postulated some additional cosmological device to carry the sun, moon, and planets around in the space that lay between them. Therefore the two-sphere universe is not really a cosmology at all, but only the structural framework for one. Furthermore, that structural framework housed a great many different and controversial astronomical and cosmological schemes during the nineteen hundred years that separate the fourth century B.C. from the age of Copernicus. There were many two-sphere universes. But, after its first establishment, the two-sphere framework itself was almost never questioned. For very nearly two millennia it guided the imagination of all astronomers and most philosophers. That is why we begin our discussion of the main Western astronomical tradition by considering the two-sphere universe, framework though it is, in abstraction from the various planetary devices advanced by one astronomer or another to complete it.

The origin of the two-sphere framework is obscure, but the source of its persuasiveness is not. The sphere of the heavens is only a short step from the domed heaven of the Egyptians and Babylonians, and the heavens do look domed. The elongation that the Egyptians gave to the heavens vanishes in a society not based upon a river like the Nile and leaves a hemispherical shell. Connecting the vault above the earth with a symmetric vault below gives the universe an appropriate and satisfying closure. The rotation of the resulting sphere is indicated by the stars themselves; as we shall shortly see, a steady rotation of the outer sphere, once every 23 hours 56 minutes, will produce just the diurnal circles that we have already described.

There is, in addition, an essentially aesthetic argument in favor of the spherical universe. Since the stars seem as far away as anything we can see and since they all move together, it is natural to suppose that they are simply markings on the outer surface of the universe and that they move with it. Furthermore, since the stars move eternally with perfect regularity, the surface on which they move ought itself be perfectly regular and it should move in the same manner forever. What figure better fulfills these conditions than the sphere, the only completely symmetric surface and one of the few that can turn eternally upon itself, occupying exactly the same space in each instant of its motion? In what other form could an eternal and self-sufficient universe be created? This is essentially the argument employed by the Greek philosopher Plato (fourth century B.C.) in his *Timaeus*, an allegorical story of the creation in which the universe appears as an organism, an animal.

[The Creator's] intention was, in the first place, that the animal should be as far as possible a perfect whole and of perfect parts: secondly, that it should be one, leaving no remnants out of which another such world might be created: and also that it should be free from old age [eternal] and unaffected by disease [incorruptible] . . . . Wherefore he made the world in the form of a globe, round as from a lathe, having its extremes in every direction equidistant from the center, the most perfect and the most like itself of all figures; for he considered that the like is infinitely fairer than the unlike. This he finished off, making the surface smooth all round for many reasons; in the first place, because the living being had no need of eyes when there was nothing remaining outside him to be seen; nor of ears when there was nothing to be heard; and there was no surrounding atmosphere to be breathed; nor would there have been any use of organs by the help of which he might receive his food or get rid of what he had already digested, since there was nothing which went from him or came into him: for there was nothing beside him. Of design he was created thus, his own waste providing his own food, and all that he did or suffered taking place in and by himself. For the Creator conceived that a being which was self-sufficient would be far more excellent than one which lacked anything; and, as he had no need to take anything or defend himself against any one, the Creator did not think it necessary to bestow upon him hands: nor had he any need of feet, nor of the whole apparatus of walking; but the movement suited to his spherical form was assigned to him, . . . and he was made to move in the same manner and on the same spot, within his own limits revolving in a circle.[2]

Some of the ancient arguments for the sphericity of the earth are of the same sort: what is more fitting than that the earth, man's abode, should display the same perfect figure with which the universe was created? But many of the demonstrations are more concrete and familiar. The hull of a ship sailing from shore disappears before the top of the mast. More of the ship and of the sea is visible from a high observation point than from a low one (Figure 10). The shadow of the earth on the moon during a lunar eclipse always has a circular edge. (This explanation of eclipses, current even before the fourth century B.C., is discussed in Section 3 of the Technical Appendix.) These arguments are still difficult to evade or refute, and in antiquity their effectiveness extended by analogy from the earth to the heavens: a celestial region that mirrored the shape of the earth seemed specially appropriate. Other arguments derived from the similarity and symmetric arrangement of the two spheres. The earth's central position, for example, kept it stationary in the spherical universe. In which direction can a body fall from the center of a sphere? There is no "down" at the

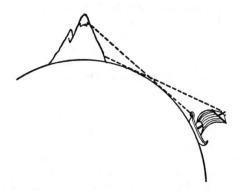

Figure 10. An ancient (and modern) argument for the earth's sphericity. An observer at the base of the mountain can see only the tip of the ship's mast over the earth's bulge. From the mountain top the entire mast and part of the hull are visible.

center, and every direction is equally "up." Therefore the earth must hang at the center, eternally stable as the universe rotates about it.

Though these arguments from symmetry may seem strange today (arguments for a discredited conclusion usually do seem strange), they were very important in ancient, medieval, and early modern thought. A discussion of symmetry, like Plato's, displays the appropriateness of the two-sphere cosmology; it explains why the universe was created in the spherical form. Even more important, as we shall discover in Chapters 3 and 4, the symmetry of the two spheres provided important links between astronomical, physical, and theological thought, because it was essential to each. In Chapter 5 we shall find Copernicus struggling vainly to preserve the essential symmetry of ancient cosmology in a universe constructed to contain a moving planetary earth. But we are now most concerned with the astronomical functions of the two-sphere universe, and here the case is entirely clear. In astronomy the two-sphere cosmology works and works very well. That is, it accounts precisely for the observations of the heavens described in the earlier portions of this chapter.

Figure 11 shows a spherical earth, much exaggerated in size, at the center of a larger sphere of the stars. An observer on the earth, at a position indicated by the arrow, O, can see just half of the sphere. His horizon is bounded by a plane (shaded in the diagram) tangent to the earth at the point where he is standing. If the earth is very small

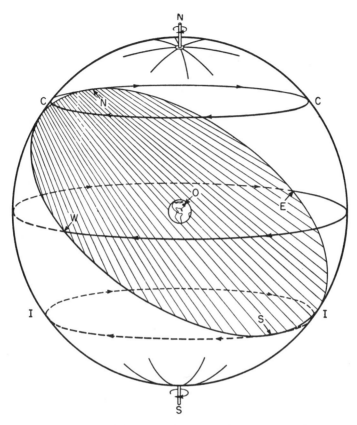

Figure 11. Astronomical functions of the two-sphere universe. The outermost circle is a cross section of the sphere of the stars which rotates steadily eastward about the axis NS. The observer at O can see all portions of this sphere that lie above the shaded horizon plane SWNE. If the diagram were drawn to scale, the earth would be much reduced and the horizon plane would be tangent to the earth at the observation point. But since a scale drawing would reduce the earth to minuscule dimensions, the plane is here drawn through the center of the sphere of the stars, and its orientation with respect to the observer is preserved by keeping it perpendicular to the line from the observer to the earth's center.

The horizontal circles in the diagram are the paths traced by selected points on the sphere as the sphere goes through its daily rotation. They are therefore the diurnal circles of selected stars, drawn solid where they are visible to the observer and broken where they lie below the horizon. The central circle is traced by a star on the celestial equator. It rises at E, due east of the observer, moves upward along a line slanting south, and so on. The uppermost and lowermost circles are traced by stars that meet the horizon in only one point. The upper circle, CC, is the diurnal circle of the southernmost circumpolar star; the lower circle, II, is traced by the northernmost of the stars that remain invisible to the observer at O.

compared with the sphere of the stars, this tangent plane will divide the outer sphere into two almost precisely equal parts, one visible to the observer, and the other hidden from him by the surface of the earth. Any objects permanently mounted on the outer sphere will, like the stars, retain the same relative positions when seen from the tiny central earth. If the sphere turns steadily about an axis through the diametrically opposite points N and S, all the stars will move with it unless they are actually located at N or S. Since S is invisible to the observer in the diagram, N is the single stationary point in his heaven, his celestial pole, and it is in fact located just about 45° over the due-north point on his horizon, as it should be for an observer at O, a point in middle-northern latitudes.

Objects near the point N on the outer sphere appear to the observer at O to rotate slowly in circles about the pole; if the sphere rotates once in every 23 hours 56 minutes, these objects complete their circles in the same period as the stars; they represent stars in the model. All stars that lie near enough the pole to be inside the circle CC on the diagram are circumpolar, for the sphere's rotation never carries them below the horizon. Stars that lie farther from N, between the circles CC and II, rise and set at an angle to the horizon once in each rotation of the sphere, but stars nearest the circle II barely get above the southern horizon and are visible only briefly. Finally, stars located within the circle II, near S, never appear to the observer at O; they are always hidden by his horizon. They would, however, be visible to an observer at other locations on the inner sphere; S is at least a potentially visible fixed point in the heavens, a second pole. Call it the south celestial pole and the visible point at N the north celestial pole.

If the observer in the diagram moves northward from O (that is, toward a point on the inner sphere directly under the north celestial pole), his horizon plane must move with him, becoming more and more nearly perpendicular to the axis of the stellar sphere as he approaches the terrestrial pole. Therefore, as the observer moves north, the celestial pole must appear farther and farther from the north point on the horizon until at last it lies directly over his head. Simultaneously, the circle CC, always drawn tangent to the northernmost point on the horizon, must expand so that more and more stars become circumpolar. Since the circle II also expands when the observer moves north, the number of invisible stars must increase as well. If the observer moves south, the effects are exactly reversed, the pole coming closer and closer to the north point on the horizon and the circles *CC* and *II* shrinking in size until they just enclose the north and south celestial poles when the observer gets to the equator. Figure 12 shows the two limiting cases, an observer at the north pole of the earth and an observer at the terrestrial equator. In the first case the horizon is shown horizontal; the north celestial pole is directly over the observer's head;

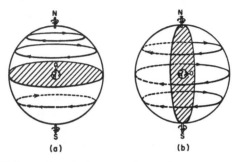

Figure 12. Stellar motions in the two-sphere universe as seen by an observer (*a*) at the north terrestrial pole and (*b*) at the equator.

the stars in the upper half of the sphere swing continuously in circles parallel to the horizon; and the stars in the lower hemisphere are never seen at all. In the second diagram the horizon is vertical; the north and south celestial poles are fixed at the north and south points on the horizon; all stars can be seen at some time or other; but none can be seen through more than a semicircle.

Except that these last extreme examples were not observed in antiquity, the motion of the stars in the two-sphere model of the universe coincides precisely with the previously discussed observations of real stars. There is no more convincing argument for the two-sphere cosmology.

The Sun in the Two-Sphere Universe

A complete discussion of the sun's motion in the two-sphere universe demands that the cosmology be elaborated to account for the sun's intermediate position between the central earth and the peripheral rotating stellar sphere. This elaboration is part of the larger problem of the planets; it will be considered in the next chapter. But

even the skeletal cosmology described above permits a great simplification in the description of the sun's apparent motion. Seen from the
central earth against the spherical backdrop provided by the stellar
sphere, the sun's motion acquires a regularity that was not apparent
until the stars were localized on a rotating earth-centered sphere.

The new simplicity of the sun's apparent motion is described in
Figure 13, which is a reduced sketch of the stellar sphere with its north
pole visible and with the direction of westward diurnal rotation indicated by an arrow about the pole. Halfway between the north and
south celestial poles is drawn the celestial equator, a great circle on
which lie all the stars (and all points on the sphere) that rise and set
exactly due east and west. A great circle is the simplest of all curves
that can be drawn on the surface of a sphere — the intersection of the
sphere's surface with that of a plane through the sphere's center —
and the new simplicity of the sun's motion results from the fact that on
a celestial sphere the ecliptic, too, is just a great circle, dividing the
sphere into two equal halves. In Figure 13 the ecliptic is the slanted

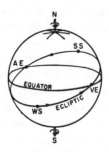

Figure 13. The equator and the ecliptic on the celestial sphere.

circle, intersecting the celestial equator in two diametrically opposite
points at an angle of 23½°; it contains all the points at which the center
of the sun is seen against the sphere of the stars by an observer on the
earth. At each instant the sun's center appears at a point on this great
circle, participating in the diurnal westward motion of the whole
sphere, but simultaneously the sun slips slowly eastward (arrows in
diagram), completing its journey around the ecliptic once in a year.

Because, during any 24-hour period, the sun appears to remain very
near a single point on the ecliptic, it must move daily in a diurnal circle

very much like that of a star. But the sun moves slowly eastward with respect to the sphere as the sphere itself turns rapidly westward. Therefore the sun must complete its diurnal circle slightly more slowly than the stars complete theirs, losing a little ground in its race with the stars each day and being completely "lapped" by them once each year. More precisely, since the sun must move through 360° to traverse the ecliptic and since it completes this journey in just over 365 days, its eastward motion along the ecliptic must cover just under 1° per day, and this is the figure derived earlier from observation (p. 23). It is the distance which the sun slips backward (or loses) with respect to the stars each day. Furthermore, since the length of the day is defined by the diurnal motion of the sun and since the stars (moving 15° per hour or 1° every 4 minutes) get 1° farther ahead of the sun each day, a star that gets, say, overhead at midnight tonight will complete its diurnal motion and return to the same position in the sky just 4 minutes before midnight tomorrow night. Once again a detail about the behavior of the heavens, initially introduced as one among many assorted observations (p. 16), has become part of a coherent pattern in the two-sphere universe.

A similar order is apparent in the positions that the equinoxes and solstices assume on the stellar sphere. The two equinoxes must be the two diametrically opposite positions on the stellar sphere where the ecliptic intersects the celestial equator. These are the only points on the ecliptic that always rise and set due east and west. Similarly, the two solstices must be the points on the ecliptic midway between the two equinoxes, for these are the points on the ecliptic that lie farthest north and south of the celestial equator. When the sun is at one of these points it must rise farther north (or south) of the due-east point than it does at any other time. Since the sun moves steadily eastward from the summer solstice toward the autumnal equinox, the individual equinoxes and solstices are readily identified on the sphere. Each of them is labeled on the ecliptic in Figure 13, and once the ecliptic has been drawn and labeled in this manner it is possible, by constructing an appropriate horizon plane inside the stellar sphere, to discover how the sun's behavior varies during the course of a year when observed from any location on the earth's surface. Three particularly significant examples of the sun's motion at various seasons of the year are derived from the two-sphere conceptual scheme in the diagrams of Figure 14. In these diagrams the full force of the conceptual scheme begins to appear.

The Functions of a Conceptual Scheme

Unlike the observations described in the early sections of this chapter, the two-sphere universe is a product of the human imagination. It is a conceptual scheme, a theory, deriving from observations but simultaneously transcending them. Because it will not yet account

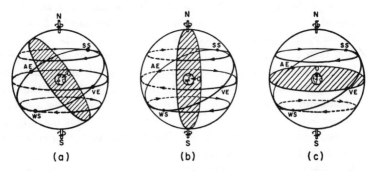

Figure 14. The motion of the sun observed from different locations on the earth.

(*a*) Observer in middle-northern latitudes: At the summer solstice the sun rises along a slanting line far north of the due-east point; more than one-half of its diurnal circle lies above the horizon, so that the days are longer than the nights. At the equinoxes the sun rises due east, and only one-half of its diurnal circle is visible. At the winter solstice the sun rises far to the south, and the days are shorter than the nights. The sun's maximum daily elevation above the horizon is greatest in summer, but at all seasons the sun's noon shadow must point due north.

(*b*) Observer at the equator: Whatever the sun's location on the ecliptic, the horizon plane divides the sun's diurnal circle into two equal parts. Days and nights are always of equal length, and there is little seasonal variation of climate. During half the year (vernal equinox to autumnal equinox) the sun rises north of the due-east point, and its noon shadow points due south. During the other half year the sun rises south of east, and its noon shadow points north.

(*c*) Observer at the north terrestrial pole: Half of the ecliptic is always below the horizon, so that for half the year (autumnal to vernal equinox) the sun is completely invisible. At the vernal equinox the sun begins to peek above the horizon, swinging around it daily in a gradually rising spiral until the summer solstice. Then the sun gradually spirals back to the horizon, disappearing slowly below it at the autumnal equinox. Between the vernal and autumnal equinoxes the sun does not set.

for the motions of all celestial bodies (the planets, in particular, have been ignored to this point), the two-sphere cosmology is not complete. But already it provides cogent illustrations of some logical and psychological functions that scientific theories can perform for the men who develop or make use of them. The evolution of any scientific conceptual scheme, astronomical or nonastronomical, depends upon the way in which it performs these functions. By making some of them explicit before the two-sphere universe is elaborated in the next two chapters, we can highlight in advance a few of the most fundamental problems that will emerge during this study of the Copernican Revolution.

Perhaps the most striking characteristic of the two-sphere universe is the assistance that it gives to the astronomer's memory. This characteristic of a conceptual scheme is often called conceptual economy. Though they were both carefully selected and systematically presented, the observations of the sun and stars discussed in earlier sections were, as a group, extremely complex. To a man not already thoroughly acquainted with the heavens, one observation, like the direction of the slant line along which the sun rises or the corresponding behavior of the gnomon's shadow, seems unrelated to another observation, like the location of the celestial pole or the brief appearance of the stars in the southern sky. Each observation is a separate item in a long list of bare facts about the heavens, and it is difficult to retain the whole list in memory simultaneously.

The two-sphere universe presents no such problem: a gigantic sphere bearing the stars rotates steadily westward on a fixed axis once every 23 hours 56 minutes; the ecliptic is a great circle on this sphere tilted 23½° to the celestial equator, and the sun moves steadily eastward about the ecliptic once every 365¼ days; the sun and stars are observed from a tiny fixed sphere located at the center of the giant stellar sphere. That much can be committed to memory once and for all, and while it is remembered the list of observations may be forgotten. The model replaces the list, because, as we have already seen, the observations can be derived from the model. Frequently they need not even be derived. A man who observes the heavens with the two-sphere universe firmly fixed in his mind will find that the conceptual scheme discloses a pattern among otherwise unrelated observations, that a list of the observations becomes a coherent whole for the first time, and that the individual items on the list are therefore more easily remembered. Without these ordered summaries which its theories provide science would be unable to accumulate such immense stores of detailed information about nature.

Because it provides a compact summary of a vast quantity of important observational materials, the two-sphere universe is actively employed by many people today. The theory and practice of both navigation and surveying can be developed with great simplicity and precision from models built to the specifications of Figure 11, and, since the model demanded by modern astronomy is far more complex, the two-sphere universe is normally used in preference to the Copernican when teaching these subjects. Most handbooks of navigation or surveying open with some sentence like this: "For present purposes we shall assume that the earth is a small stationary sphere whose center coincides with that of a much larger rotating stellar sphere." Evaluated in terms of economy, the two-sphere universe therefore remains what it has always been: an extremely successful theory.

In other respects, however, the two-sphere universe is no longer at all successful and has not been since the Copernican Revolution. It has remained economical only because economy is a purely logical function. The celestial observations known to ancient astronomers and used by modern navigators are logical consequences of the two-sphere model whether or not the model is thought to represent reality. The attitude of the scientist, his belief in the "truth" of the conceptual scheme, does not affect the scheme's logical ability to provide an economical summary. But conceptual schemes have psychological as well as logical functions, and these do depend upon the scientist's belief or incredulity. For example, the psychological craving for at-homeness, discussed in the second section, can be satisfied by a conceptual scheme only if that scheme is thought to be more than a convenient device for summarizing what is already known. During antiquity and again in the later Middle Ages the European world did have this additional commitment to the conception of a two-sphere universe. Scientists and nonscientists alike believed that the stars really were bright spots on a gigantic sphere that symmetrically enclosed man's terrestrial abode. As a result, two-sphere cosmology did for centuries provide many men with a world view, defining their place in the created world and giving physical meaning to their relation with the gods. As we shall see in

Chapters 3 and 4, a conceptual scheme that is believed and that therefore functions as part of a cosmology has more than scientific significance.

Belief also affects the way in which conceptual schemes function
within the sciences. Economy as a purely logical function, and cosmological satisfaction as a purely psychological function, lie at opposite
ends of a spectrum. Many other significant functions lie within the
spectrum, between these limits, depending both upon the logical structure of the theory and upon its psychological appeal, its ability to
evoke belief. For example, an astronomer who believes in the validity
of the two-sphere universe will find that the theory not only provides
a convenient summary of the appearances, but that it also *explains*
them, enabling him to *understand* why they are what they are. Words
like "explain" and "understand" apparently refer simultaneously to the
logical and psychological aspects of conceptual schemes. Logically,
the two-sphere universe explains the motions of the stars because the
motions can be deduced from the far simpler model. Complexity is
reduced, and such logical reduction is one essential component of
explanation. But it is not the only one. Psychologically, the two-sphere
universe provides no explanation unless it is believed to be true. The
modern navigator uses the two-sphere universe on his job, but he does
not explain the stellar motions in terms of a rotation of the outer sphere.
He believes that the diurnal motion of the stars is only an apparent
motion, and he must therefore explain it as the result of a real rotation of the earth.

A scientist's willingness to use a conceptual scheme in explanations is an index of his commitment to the scheme, a token of his belief that his model is the only valid one. Such commitment or belief
is always rash, because economy and cosmological satisfaction cannot
guarantee truth, whatever "truth" may mean. The history of science is
cluttered with the relics of conceptual schemes that were once fer-
vently believed and that have since been replaced by incompatible
theories. There is no way of proving that a conceptual scheme is final.
But, rash or not, this commitment to a conceptual scheme is a common
phenomenon in the sciences, and it seems an indispensable one, because it endows conceptual schemes with one new and all-important
function. Conceptual schemes are comprehensive; their consequences
are not limited to what is already known. Therefore, an astronomer

committed to, say, the two-sphere universe will expect nature to show the additional, but as yet unobserved, properties that the conceptual scheme predicts. For him the theory will transcend the known, becoming first and foremost a powerful tool for predicting and exploring the unknown. It will affect the future of science as well as its past.

The two-sphere universe tells the scientist about the behavior of the sun and stars in parts of the world (like the southern hemisphere and the terrestrial poles) to which he has never traveled. In addition it informs him of the motion of stars that he has never observed systematically. Since they are fastened to the stellar sphere, they must revolve in diurnal circles as the other stars do. This is new knowledge, derived initially not from observation but directly from the conceptual scheme, and such new knowledge can be immensely consequential. For example, two-sphere cosmology states that the earth has a circumference, and it suggests a set of observations (discussed in Section 4 of the Technical Appendix) by which the astronomer can discover how large the earth's circumference is. One set of these observations (a bad one, as it happened, for the resulting value of the circumference was far too small) led Columbus to believe that the circumnavigation of the globe was a practical undertaking, and the results of his voyages have been recorded. Those voyages and the subsequent travels of Magellan and others provided observational evidence for beliefs that had previously been derived solely from theory, and they supplied science with many unanticipated observations besides. The voyages would not have been undertaken, and the novel observations would not have accrued to the sciences, if a conceptual scheme had not pointed the way.

Columbus' voyages are one example of the fruitfulness of a conceptual scheme. They show how theories can guide a scientist into the unknown, telling him where to look and what he may expect to find, and this is perhaps the single most important function of conceptual schemes in science. But the guidance provided by conceptual schemes is rarely so direct and unequivocal as that illustrated above. Typically a conceptual scheme provides hints for the organization of research rather than explicit directives, and the pursuit of these hints usually requires extension or modification of the conceptual scheme that provided them. For example, the two-sphere universe was initially developed principally to account for the diurnal motions of the stars and for the way in which those motions varied with the observer's location on the earth. But once it had been developed, the new theory was readily extended to give order and simplicity to observations of the sun's motion as well. And, having disclosed the unsuspected regularity that underlay the complexity of the sun's behavior, the conceptual scheme provided a framework within which could be studied the even more irregular motions of the planets. That problem had seemed unmanageable until the over-all motion of the heavens was reduced to order.

Much of this book will be concerned with the fruitfulness of particular conceptual schemes, that is, with their effectiveness as guides for research and as frameworks for the organization of knowledge. The next two chapters, in particular, will examine the role of the two-sphere universe in the ancient solution, first, of the problem of the planets and, then, of some problems lying entirely outside astronomy. Later we shall discover the rather different sort of guidance given to scientific research by Copernicus' novel conception of a moving planetary earth. The very best example of fruitfulness is, however, the story told in the whole of this book. The Copernican universe is itself the product of a series of investigations that the two-sphere universe made possible: the conception of a planetary earth is the most forceful illustration of the effective guidance given to science by the incompatible conception of a unique central earth. That is why a discussion of the Copernican Revolution must begin with a study of the two-sphere cosmology which Copernicanism ultimately made obsolete. The two-sphere universe is the parent of the Copernican; no conceptual scheme is born from nothing.

Ancient Competitors of the Two-Sphere Universe

The two-sphere conception of the universe was not the only cosmology suggested in ancient Greece. But it was the one taken most seriously by the largest number of people, particularly by astronomers, and it was the one that later Western civilization first inherited from the Greeks. Yet many of the alternate cosmologies proposed and rejected in antiquity show far closer resemblances to modern cosmological beliefs than does the two-sphere universe. Nothing more clearly illustrates the strengths of the two-sphere cosmology and foreshadows the difficulties to be encountered in overthrowing it than a comparison of the scheme with a few of its superficially more modern alternatives.

As early as the fifth century B.C., the Greek atomists, Leucippus and Democritus, visualized the universe as an infinite empty space, populated by an infinite number of minute indivisible particles or atoms moving in all directions. In their universe the earth was but one of many essentially similar heavenly bodies formed by the chance aggregation of atoms. It was not unique, nor at rest, nor at the center. In fact, an infinite universe has no center; each part of space is like every other; therefore the infinite number of atoms, some of which aggregated to form our earth and sun, must have formed numerous other worlds in other portions of the empty space or void. For the atomists there were other suns and other earths among the stars.

Later in the fifth century the followers of Pythagoras suggested a second cosmology which set the earth in motion and partially deprived it of its unique status. The Pythagoreans did place the stars on a gigantic moving sphere, but at the center of this sphere they placed an immense fire, the Altar of Zeus, invisible from the earth. The fire could not be seen, because the earth's populated areas were always turned away from the fire. For the Pythagoreans the earth was just one of a number of celestial bodies, including the sun, all of which moved in circles about the central fire. A century later Heraclides of Pontus (fourth century B.C.) suggested that it was a daily rotation of the central earth rather than a rotation of the peripheral sphere of the stars that produced the apparent motion of the heavens. He also obscured the symmetry of the two-sphere universe by suggesting that the planets Mercury and Venus revolved in circles about the moving sun rather than in independent circular orbits about the central earth (see Chapter 2). Still later, in the middle of the third century B.C., Aristarchus of Samos, whose ingenious and influential measurements of astronomical dimensions are described in the Technical Appendix, advanced the proposal that has earned for him the title of "the Copernicus of antiquity." For Aristarchus the sun was at the center of an immensely expanded sphere of the stars, and the earth moved in a circle about the sun.

These alternative cosmologies, particularly the first and last, are remarkably like our modern views. We do believe today that the earth is but one of a number of planets, circulating about the sun, and that the sun is but one of a multitude of stars, some of which may have their own planets. But though some of these speculative suggestions gave rise to significant minority traditions in antiquity, and though all of them were a continuing source of intellectual stimulus to innovators like Copernicus, they were not originally supported by the arguments that now make us believe them, and in the absence of these arguments they were rejected by most philosophers and almost all astronomers in the ancient world. In the Middle Ages they were ridiculed or ignored. The reasons for the rejection were excellent. These alternative cosmologies violate the first and most fundamental suggestions provided by the senses about the structure of the universe. Furthermore, this violation of common sense is not compensated for by any increase in the effectiveness with which they account for the appearances. At best they are no more economical, fruitful, or precise than the two-sphere universe, and they are a great deal harder to believe. It was difficult to take them seriously as explanations.

All of these alternative cosmologies take the motion of the earth as a premise, and all (except Heraclides' system) make the earth move as one of a number of heavenly bodies. But the first distinction suggested by the senses is that separating the earth and the heavens. The earth is not part of the heavens; it is the platform from which we view them. And the platform shares few or no apparent characteristics with the celestial bodies seen from it. The heavenly bodies seem bright points of light, the earth an immense nonluminous sphere of mud and rock. Little change is observed in the heavens: the stars are the same night after night and apparently have remained so throughout the many centuries covered by ancient records. In contrast the earth is the home of birth and change and destruction. Vegetation and animals alter from week to week; civilizations rise and fall from century to century; legends attest the slower topographical changes produced on earth by flood and storm. It seems absurd to make the earth like celestial bodies whose most prominent characteristic is that immutable regularity never to be achieved on the corruptible earth.

The idea that the earth moves seems initially equally absurd. Our senses tell us all we know of motion, and they indicate no motion for the earth. Until it is reëducated, common sense tells us that, if the earth is in motion, then the air, clouds, birds, and other objects not attached to the earth must be left behind. A man jumping would descend to earth far from the point where his leap began, for the earth would move beneath him while he was in the air. Rocks and trees, cows and men must be hurled from a rotating earth as a stone flies from a rotating sling. Since none of these effects is seen, the earth is at rest. Observation and reason have combined to prove it.

Today in the Western world only children argue this way, and only children believe that the earth is at rest. At an early age the authority of teachers, parents, and texts persuades them that the earth is really a planet and in motion; their common sense is reëducated; and the arguments born from everyday experience lose their force. But reëducation is essential — in its absence these arguments are immensely persuasive — and the pedagogic authorities that we and our children accept were not available to the ancients. The Greeks could only rely on observation and reason, and neither produced evidence for the earth's motion. Without the aid of telescopes or of elaborate mathematical arguments that have no apparent relation to astronomy, no effective evidence for a moving planetary earth can be produced. The observations available to the naked eye fit the two-sphere universe very well (remember the universe of the practical navigator and surveyor), and there is no more natural explanation of them. It is not hard to realize why the ancients believed in the two-sphere universe. The problem is to discover why the conception was given up.

2

THE PROBLEM

OF THE PLANETS

Apparent Planetary Motion

If the sun and stars were the only celestial bodies visible to the naked eye, modern man might still accept the fundamental tenets of the two-sphere universe. Certainly he would have accepted them until the invention of the telescope, more than half a century after Copernicus' death. There are, however, other prominent celestial bodies, particularly the planets, and the astronomer's interest in these bodies is the principal source of the Copernican Revolution. Once again we consider observations before dealing with interpretive explanations. And once again the discussion of interpretations will confront us with a new and fundamental problem about the anatomy of scientific belief.

The term planet is derived from a Greek word meaning "wanderer," and it was employed until after Copernicus' lifetime to distinguish those celestial bodies that moved or "wandered" among the stars from those whose relative positions were fixed. For the Greeks and their successors the sun was one of the seven planets. The others were the moon, Mercury, Venus, Mars, Jupiter, and Saturn. The stars and these seven planets were the only bodies recognized as celestial in antiquity. No additional planets were discovered until 1781, long after the Copernican theory had been accepted. Comets, which were well known in the ancient world, were not considered celestial bodies before the Copernican Revolution (Chapter 6).

All of the planets behave somewhat like the sun, though their motions are uniformly more complex. All have a westward diurnal motion with the stars, and all move gradually eastward among the stars until they return to approximately their original positions. Throughout their motions the planets stay near the ecliptic, sometimes wandering north of it, sometimes south, but very seldom leaving the band of the zodiac, an imaginary strip in the sky extending for 8° on either side of the ecliptic. At this point the resemblance between planets ends, and the study of planetary irregularities begins.

The moon travels around the ecliptic faster and less steadily than the sun. On the average it completes one journey through the zodiac in 27⅓ days, but the time required for any single journey may differ from the average by as much as 7 hours. In addition, the appearance of the moon's disk changes markedly as it moves. At new moon its disk is completely invisible or very dim; then a thin bright crescent appears, which gradually waxes until, about a week after new moon, a semi-circular sector is visible. About 2 weeks after new moon the full circular disk appears; then the cycle of phases is reversed, and the moon gradually wanes, reaching new moon again about 1 month after the preceding new moon. The cycle of phases is recurrent, like the moon's journey through the signs of the zodiac, but the two lunar cycles are significantly out of step. New moon recurs after an average interval of 29½ days (individual cycles may differ by as much as ½ day from this average), and, since this is 2 days longer than the period of an average journey around the zodiac, the position of successive new moons must gradually move eastward through the constellations. If new moon occurs at the position of the vernal equinox one month, the moon will still be waning when it returns to the vernal equinox 27⅓ days later. New moon does not recur for about 2 days more, by which time the moon has moved almost 30° east from the equinox.

Because they are easily visible and conveniently spaced, the moon's phases provided the oldest of all calendar units. Primitive forms of both the week and the month appear in a Babylonian calendar from the third millennium B.C., a calendar in which each month began with the first appearance of the crescent moon and was subdivided at the 7th, 14th, and 21st days by the recurrent "quarters" of the moon's cycle. At the dawn of civilization men must have counted new moons and quarters to measure time intervals, and as civilization progressed they repeatedly attempted to organize these fundamental units into a co-herent long-term calendar — one that would permit the compilation of historical records and the preparation of contracts to be honored at a specified future date.

But at this point the simple obvious lunar unit proved intractable. Successive new moons may be separated by intervals of either 29 or 30 days, and only a complex mathematical theory, demanding generations of systematic observation and study, can determine the length of a specified future month. Other difficulties derive from the incommensurable lengths of the average lunar and solar cycles. Most societies (but not all, for pure lunar calendars are still used in parts of the Middle East) must adjust their calendars to the sun-governed annual climatic variation, and for this purpose some systematic method for inserting an occasional thirteenth month into a basic year of 12 lunar months (354 days) must be devised. These seem to have been the first difficult technical problems encountered by ancient astronomy. More than any others, they are responsible for the birth of quantitative planetary observation and theory. The Babylonian astronomers who finally solved these difficulties between the eighth and third centuries B.C., a period during most of which Greek science was still in its infancy, accumulated much of the fundamental data subsequently incorporated into the developed structure of the two-sphere universe.

Unlike the moon and sun, the remaining five planets appear as mere points of light in the heavens. The untrained naked-eye observer can distinguish them from stars with assurance only by a series of observations that discloses their gradual motion around the ecliptic. Usually the planets move eastward through the constellations: this is their so-called "normal motion." On the average, both Mercury and Venus require 1 year for each complete circuit of the zodiac; the length of Mars's cycle averages 687 days; Jupiter's average period is 12 years; and Saturn's is 29 years. But in all cases the time required for any single journey may be quite different from the average period. Even when moving eastward through the stars, a planet does not continue at a uniform rate.

Nor is its motion uniformly eastward. The normal motion of all planets except the sun and moon is occasionally interrupted by brief intervals of westward or "retrograde" motion. Compare Mars retrogressing in the constellation Taurus, shown in Figure 15, with the normal motion of the sun through Taurus, shown in Figure 9 (p. 22). Mars enters the diagram in normal (eastward) motion, but as its motion continues, the planet gradually slows until at last it reverses its direction and begins to move westward, in retrograde. Other planets behave in much the same way, each one repeating the interlude of retrograde motion after a fixed length of time. Mercury briefly reverses its motion through the stars once every 116 days, and Venus retrogresses every 584 days. Mars, Jupiter, and Saturn show retrograde motion every 780, 399, and 378 days, respectively.

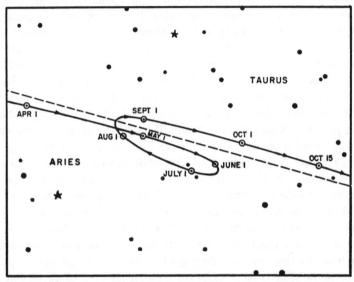

Figure 15. Mars retrogressing in Aries and Taurus. The section of sky is the same as that shown in Figure 9 and in the box on the star map of Figure 8. The broken line is the ecliptic and the solid line the path of the planet. Note that Mars does not stay on the ecliptic and that, though its over-all motion is eastward among the stars, there is a period from the middle of June to early August during which it moves to the west. The retrogressions of Mars are always of approximately this form and duration, but they do not always occur on the same date or in the same part of the sky.

In their gradual eastward motions interrupted by periodic westward retrogressions, the five wandering stars behave quite similarly. But there is an additional characteristic of their motion which divides them into two groups; this is the correlation between their position and the sun's. Mercury and Venus, the two so-called inferior planets, never get very far from the sun. Mercury is always found within 28° of the sun's moving disk, and Venus's maximum "elongation" is 45°. Both planets move in a continuous slow shuttle, back and forth across the moving sun; for a time they move eastward with the sun, then retrogress across its disk, and finally reverse themselves to overtake the sun once more. When to the east of the sun, either of these inferior planets appears as an "evening star," becoming visible shortly after sunset and then rapidly following the sun below the horizon. After retrogressing westward across the sun's disk, the planet becomes a "morning star," rising shortly before dawn and disappearing in the brilliant light of sunrise. But in between, when close to the sun, neither Mercury nor Venus can be seen at all. Therefore, until their motion was analyzed with respect to the sphere of the stars, neither of the inferior planets was recognized as the same celestial body when it appeared as a morning and as an evening star. For millenniums Venus had one name when it rose in the east shortly before dawn and another when, weeks later, it again became visible just over the western horizon shortly after sunset.

Unlike Mercury and Venus, the superior planets, Mars, Jupiter, and Saturn, are not restricted to the same part of the sky as the sun. Sometimes they are very close to or "in conjunction" with it; at other times they are 180° across the sky or "in opposition" to the sun; between these times they assume all the intermediate positions. But though their positions are unrestricted, their behavior does depend upon their relation to the sun. Superior planets retrogress only when they are in opposition. Also, when in retrograde motion across the sky from the sun, superior planets appear brighter than at any other time. This increased brilliance, which has usually been interpreted (at least since the fourth century B.C.) as indicating a decrease in the planet's distance from the earth, is particularly striking in the case of Mars. Normally a relatively inconspicuous planet, Mars in opposition will frequently outshine every celestial body in the night sky except the moon and Venus.

Interest in the five wandering stars is by no means so ancient as a concern with the sun and moon, presumably because the wandering stars had no obvious practical bearing upon the lives of ancient peoples. Yet observations of the appearance and disappearance of Venus were recorded in Mesopotamia as early as 1900 B.C., probably as omens, portents of the future, like the signs to be read in the entrails of sacrificial sheep. These scattered observations presage the much later development of systematic astrology, a means of forecasting whose intimate relation to the development of planetary astronomy is considered in the next chapter. The same concern with omens clearly motivated the more systematic and complete records of eclipses, retrograde motions, and other striking planetary phenomena compiled by Babylonian observers after the middle of the eighth century B.C. Ptolemy, the dean of ancient astronomers, later complained that even these records were fragmentary, but fragmentary or not they provided the first data capable of specifying the full-scale problem of the planets as that problem was to develop in Greece after the fourth century B.C.

The problem of the planets is partially specified by the description of the planetary motions sketched in the preceding pages. How are the complex and variable planetary motions to be reduced to a simple and recurrent order? Why do the planets retrogress, and how account for the irregular rate of even their normal motions? These questions indicate the direction of most astronomical research during the two millenniums from the time of Plato to the time of Copernicus. But because it is almost entirely qualitative, the preceding description of the planets does not specify the problem fully. It states a simplified problem and in some respects a misleading one. As we shall shortly see, qualitatively adequate planetary theories are easily invented: the description above can be reduced to order in several ways. The astronomer's problem, on the other hand, is by no means simple. He must explain not merely the existence of an intermittent westward motion superimposed upon an over-all eastward motion through the stars, but also the precise position that each planet occupies among the stars on different days, months, and years over a long period of time. The real problem of the planets, the one that leads at last to the Copernican Revolution, is the quantitative problem described in lengthy tables which specify in degrees and minutes of arc the varying position of every planet.

The Location of the Planets

The two-sphere universe, as developed in the last chapter, provided no explicit information about the positions or motions of the seven planets. Even the sun's location was not discussed. To appear "at" the vernal equinox (or any other point on the stellar sphere) the sun need merely be somewhere on a line stretching from the observer's eye to or through the appropriate point in the background of stars.

Like the other planets, it might be either inside, on, or perhaps even outside the sphere of the stars. But though the two-sphere universe fails to specify the shape or location of the planetary orbits, it does make certain choices of position and orbit more plausible than others, and it therefore at once guides and restricts the astronomer's approach to the problem of the planets. That problem was set by the results of observation, but, from the fourth century B.C., it was pursued in the conceptual climate of two-sphere cosmology. Both observation and theory made essential contributions to it.

Within a two-sphere cosmology, for example, the planetary orbits should if possible preserve and extend the fundamental symmetry embodied in the first two spheres. Ideally the orbits should therefore be earth-centered circles, and the planets should revolve in these circles with the same regularity that is exemplified in the rotation of the stellar sphere. The ideal does not quite conform to observation. As we shall see presently, an earth-centered circular orbit located in the plane of the ecliptic provides a good account of the sun's annual motion, and a similar circle can give an approximate account of the somewhat less regular motion of the moon. But circular orbits do not even hint at an explanation of the gross irregularities, like retrogression, observed in the motions of the other five wandering "stars." Nevertheless, astronomers who believed in the two-sphere universe could, and for centuries did, think that earth-centered circles were the natural orbits for planets. Such orbits at least accounted for the over-all average eastward motions. Observed deviations from the average motion — changes in the rate or direction of a planet's motion — indicated that the planet itself had deviated from its natural circular orbit, to which it would again return. On this analysis the problem of the planets became simply that of explaining the observed deviation from average motion through the stars in terms of a corresponding deviation of each planet from its single circular orbit.

We shall examine some of the ancient explanations of these deviations in the next three sections, but first notice, as the ancients also did, how far it is possible to proceed by neglecting the planetary irregularities and assuming simply that all orbits are at least approximately circular. Almost certainly, in the two-sphere universe, the planets move in the region between the earth and the stars. The stellar sphere itself was often viewed as the outer boundary of the universe, so that the planets could not be outside it; the difference between planetary and stellar motions indicated that the planets were probably not located on the sphere, but in some intermediate region where they were affected by some influence that was inactive at the stellar sphere; the whole argument gained force from the detail visible on the face of the moon, presumptive evidence that one planet, at least, must be nearer than the stars. Ancient astronomers, therefore, laid out the planetary orbits in that vast and previously empty space between the earth and the sphere of the stars. By the end of the fourth century B.C., the two-sphere universe was filling up. Later it was to become crowded.

Once the general location and shape of their orbits were known, it proved possible to make a plausible and satisfying guess about the order in which the planets were arranged. Planets like Saturn and Jupiter, whose eastward motion was slow and whose total motion, therefore, very nearly kept pace with the stars, were supposed to be close to the stellar sphere and far from the earth. The moon, on the other hand, which loses over 12° a day in its race with the stars, must be closer to the stationary surface of the earth. Some ancient philosophers seem to have justified this hypothetical arrangement by imagining that the planets floated in a gigantic aethereal vortex, whose outer surface moved rapidly with the sphere of the stars and whose interior was at rest at the earth's surface. Any planet caught in such a vortex would lose more ground with respect to the stellar sphere if it were closer to the earth. Other philosophers reached the same conclusion by a different sort of argument, later recorded, at least in its essentials, by the Roman architect Vitruvius (first century B.C.). In analyzing the differences between the intervals required by different planets for trips about the ecliptic, Vitruvius suggested an illuminating analogy:

Place seven ants on a wheel such as potters use, having made seven channels on the wheel about the center, increasing successively in circumference; and suppose those ants obliged to make a circuit in these channels while the wheel is turned in the opposite direction. In spite of having to move in a direction contrary to that of the wheel, the ants must necessarily complete their journeys in the opposite direction, and that ant which is nearest the center must finish its circuit sooner, while the ant that is going round at the outer edge of the disk of the wheel must, on account of the size of its circuit, be much slower in completing its course, even though it is moving just as quickly as the other. In the same way, these stars, which struggle on against the course of the firmament, are accomplishing an orbit on paths of their own; but, owing to the revolution of the heaven, they are swept back as it goes round every day.[1]

Before the end of the fourth century B.C., arguments like the above had led to an image of the universe similar to the one sketched in Figure 16; diagrams like this, or their verbal equivalents, remained current in elementary books on astronomy or cosmology until the early seventeenth century, long after Copernicus' death. The earth is at the center of the stellar sphere, which bounds the universe; immediately inside this outer sphere is the orbit of Saturn, the planet that takes longest to move around the zodiac; next comes Jupiter and then Mars.

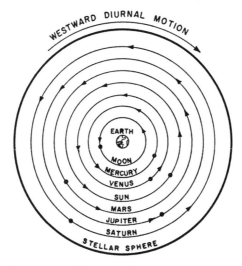

Figure 16. Approximate planetary orbits in the two-sphere universe. The outermost circle is a cross section of the stellar sphere in the plane of the ecliptic.

To this point the order is unambiguous: the planets are arranged, from the outside, in the order of decreasing orbital period; the same technique places the lunar orbit closest to the earth. But the remaining three planets present a problem; the sun, Venus, and Mercury all complete their journeys about the earth in the same average time, 1 year, and their order therefore cannot be determined by the device applied to the other planets. There was, in fact, much disagreement about their order during antiquity. Until the second century B.C. most astronomers placed the sun's orbit just outside the moon's, Venus's outside the sun's, then Mercury's, and then Mars's. After that date, however, the order shown in the diagram — moon, Mercury, Venus, sun, Mars, etc. — became increasingly popular. In particular, it was adopted by Ptolemy, and his authority imposed it upon most of his successors. We shall therefore take this order as standard throughout the early chapters of this book.

As a structural diagram Figure 16 is still very crude. It gives no meaningful indication of relative dimensions of the various orbits, and it makes no attempt to provide for the observed planetary irregularities. But the conception of the universe embodied in the diagram had two important functions in the subsequent development of astronomy and cosmology. In the first place, the diagram contains most of the structural information about the earth-centered universe that ever became common knowledge among nonastronomers. The further achievements of ancient astronomy, to which we shall turn in a moment, were too mathematical for most laymen to understand. As the next two chapters indicate more fully, the most influential cosmologies developed in antiquity and the Middle Ages did not follow ancient astronomy very far beyond this point. Astronomy now becomes esoteric; its further development does not provide man with a home.

In addition, the structural diagram in Figure 16 is, despite its crudity, an immensely powerful tool in astronomical research. For many purposes it proved both economical and fruitful. For example, during the fourth century B.C., the concepts embodied in the diagram provided a complete qualitative explanation of both the phases of the moon and lunar eclipses; during the fourth and third centuries these same concepts led to a series of relatively accurate determinations of the circumference of the earth; and during the second century B.C., they provided the basis for a brilliantly conceived estimate of the sizes and distances of the sun and moon. These explanations and measurements, particularly the last, typify the immense ingenuity and power of the ancient astronomical tradition. They are, however, here relegated to the Technical Appendix (Sections 3 and 4), because they were not affected by the change in astronomical theory during the Copernican Revolution. Nevertheless, they are relevant to the Revolution. The ability of the developed two-sphere universe to explain and ultimately to predict prominent celestial phenomena like eclipses, as well as its ability to specify some linear dimensions of the celestial regions, immeasurably increased the hold of the two-sphere conceptual scheme upon the minds of both astronomers and laymen.

These achievements do not, however, touch upon the fundamental problem posed by the continuous irregularity of planetary motion, and this problem provides the pivot upon which the Copernican Revolution ultimately turns. Like so many other problems of ancient astronomy, it seems first to have emerged during the fourth century B.C., when the two-sphere universe, by explaining diurnal motion, enabled Greek astronomers to isolate the residual planetary irregularities for the first time. During the following five centuries successive attempts to explain these irregularities produced several planetary theories of unprecedented accuracy and power. But these attempts also constitute the most abstruse and mathematical part of ancient astronomy, and they are therefore usually omitted from books like this. Though a simplified précis of ancient planetary theory seems a minimal requisite for an understanding of the Copernican Revolution, a few readers may prefer to skim the next three sections (particularly the first, in which the technical presentation is especially compact), picking up the narrative again with the discussion of scientific belief that closes this chapter.

The Theory of Homocentric Spheres

The philosopher Plato, whose searching questions dominated so much of subsequent Greek thought, seems to have been the first to enunciate the problem of the planets, too. Early in the fourth century B.C. Plato is said to have asked: "What are the uniform and ordered movements by the assumption of which the apparent movements of the planets can be accounted for?" [2] and the first answer to this question was provided by his onetime pupil Eudoxus (c.408 – c.355 B.C.). In Eudoxus' planetary system each planet was placed upon the inner sphere of a group of two or more interconnected, concentric spheres whose simultaneous rotation about different axes produced the observed motion of the planet. Figure 17a shows a cross section of two such interlocked spheres whose common center is the earth and whose points of contact are the ends of the slanted axis of the inner sphere, which serve as pivots. The outer sphere is the sphere of the stars, or at least it has the same motion as that sphere; its axis passes through the north and south celestial poles, and it rotates westward about this axis once in 23 hours 56 minutes. The inner sphere's axis makes contact with the outer sphere at two diametrically opposite points 23½° away from the north and south celestial poles; therefore the equator of the inner sphere, viewed from the earth, always falls on the ecliptic of the sphere of the stars, regardless of the rotation of the two spheres.

If the sun is now placed at a point on the equator of the inner sphere and that sphere is turned slowly eastward about its axis once in a year while the outer sphere turns about its axis once a day, the sum of the two motions will reproduce the observed motion of the sun. The outer sphere produces the observed westward diurnal motion of rising and setting; the inner sphere produces the slower annual eastward motion of the sun, around the ecliptic. Similarly, if one eastward rotation of the inner sphere occurs every 27⅓ days and if the moon is placed on the equator of this sphere, then the motion of this inner sphere must produce the average motion of the moon around the ecliptic. The north and south deviations of the moon from the ecliptic and some of the irregularities in the time required by the moon for successive journeys can be approximated by adding one more very slowly moving sphere to the system. Eudoxus also used (though unnecessarily) a third sphere to describe the motion of the sun, so that six spheres were required to treat the moon and sun together.

The spheres shown in Figure 17a were known as homocentric spheres, because they have a common center, the earth. Two or three such spheres can approximately represent the total motion of the sun and of the moon, but they cannot account for the retrograde motions of the planets, and Eudoxus' greatest genius as a geometer was displayed in the modification of the system that he introduced in treating the apparent behavior of the remaining five planets. For each of these he used a total of four spheres, sketched in cross section in Figure 17b. The two outer spheres move just like the spheres of Figure 17a: the outer sphere has the diurnal motion of the sphere of the stars and the second sphere (counting from the outside) turns eastward once in the *average* time required by the planet to complete a journey around the ecliptic. (Jupiter's second sphere, for example, turns once in 12 years.) The third sphere is in contact with the second sphere at two diametrically opposite points on the ecliptic (the equator of the second sphere), and the axis of the fourth or innermost sphere is fastened to the third sphere at an angle that depends upon the characteristics of the motion to be described. The planet itself (Jupiter in the example above) is located on the equator of the fourth sphere.

Suppose now that the two outer spheres are held stationary and that the two innermost spheres rotate in opposite directions, each completing one axial rotation in the interval that separates two suc-

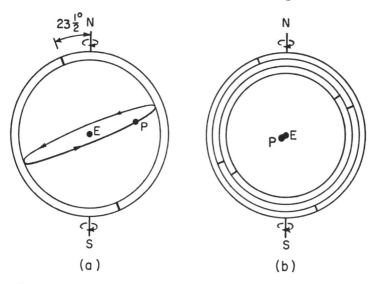

(a)   (b)

Figure 17. Homocentric spheres. In the two-sphere system (a) the outermost sphere produces the diurnal rotation, and the inner sphere moves the planet (sun or moon) steadily eastward around the ecliptic. In the four-sphere system (b) the planet P lies out of the plane of the paper, almost on a line from the earth E to the reader's eye. The two innermost spheres then produce the looped motion shown in Figure 18, and the two outer spheres produce both the diurnal motion and the average eastward planetary drift.

cessive retrogressions of the planet (399 days for Jupiter). An observer watching the motion of the planet against the temporarily stationary second sphere will see it move slowly in a figure eight, both of whose loops are bisected by the ecliptic. This is the motion sketched in Figure 18; the planet passes slowly around the loops from positions 1 to 2, 2 to 3, 3 to 4, . . . , spending equal times between each numbered point and the next, and returning to its starting point after the interval between retrogressions. During its motion from 1 to 3 to 5 the planet moves eastward along the ecliptic; during the other half of the time, while the planet moves from 5 to 7 and back to 1, it moves westward.

Now allow the second sphere to rotate eastward, carrying the two rotating inner spheres with it, and suppose that the total motion of the planet is observed against the background of stars on the first sphere, again held temporarily stationary. At all times the planet is moved eastward by the motion of the second sphere, and half of the time (while it moves from 1 to 5 in Figure 18) the planet receives an addi-

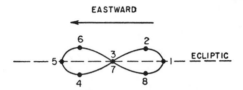

Figure 18. The looped motion generated by the two innermost homocentric spheres. In the full four-sphere system this looped motion is combined with the steady eastward motion of the second sphere, a motion that by itself would carry the planet around the ecliptic at a uniform rate. When the looped motion is added, the total motion of the planet varies in rate and is no longer confined to the ecliptic. While the planet moves from 1 to 5 on the loop, its total motion is more rapid than the average eastward motion generated by the second sphere; while the planet moves back from 5 to 1 on the loop, its eastward motion is slower than that produced by the second sphere, and when it gets near 3, it may actually move westward, in retrograde.

tional eastward motion from the two inner spheres so that its net motion is eastward and even more rapid than that of the second sphere alone. But during the other half of the time (as the planet moves from 5 to 1 in Figure 18) the eastward motion of the second sphere is opposed by a westward motion due to the two inner spheres, and, when this westward motion is most rapid (near 7 in Figure 18), the net motion of the planet against the sphere of the stars may actually be to the west, in the retrograde direction. This is just the characteristic of observed planetary motions that Eudoxus was striving to reproduce in his model.

A system of four interlocked homocentric spheres approximates the retrograde motion of Jupiter, and a second set of four spheres can account for the motion of Saturn. For each of the three remaining planets, five spheres are needed (an extension provided by Eudoxus' successor Callippus around 330 B.C.), and analysis of the resulting motions becomes correspondingly more complex. Fortunately, we need not pursue these complex combinations of rotating spheres further, because all homocentric systems have one severe drawback which in antiquity led to their early demise. Since Eudoxus' theory places each planet on a sphere concentric with the earth, the distance between a planet and the earth cannot vary. But planets appear brighter, and therefore seem closer to the earth, when they retrogress. During antiquity the homocentric system was frequently criticized for its failure to explain this variation in planetary brilliance, and the system was abandoned by most astronomers almost as soon as a more adequate explanation of the appearances was proposed.

But though short-lived as a significant astronomical device, homocentric spheres play a major role in the development of astronomical and cosmological thought. By a historical accident the century during which they seemed to provide the most promising explanation of planetary motion embraced most of the lifetime of the Greek philosopher Aristotle, who incorporated them in the most comprehensive, detailed, and influential cosmology developed in the ancient world. No comparably complete cosmology ever incorporated the mathematical system of epicycles and deferents which, in the centuries after Aristotle's death, was employed to explain planetary motion. The conception that planets are set in rotating spherical shells concentric with the earth remained an accepted portion of cosmological thought until early in the seventeenth century. Even the writings of Copernicus show important vestiges of this conception. In the title of Copernicus' great work, *De Revolutionibus Orbium Caelestium*, the "orbs" or spheres are not the planets themselves but rather the concentric spherical shells in which the planets and the stars are set.

Epicycles and Deferents

The origin of the device that replaced homocentric spheres in explaining the details of planetary motion is unknown, but its features were early investigated and developed by two Greek astronomers and mathematicians, Apollonius and Hipparchus, whose work spans the period from the middle of the third century to the end of the second century B.C. In its simplest form (Figure 19a) the new mathematical mechanism for the planets consists of a small circle, the epicycle, which rotates uniformly about a point on the circumference of a second rotating circle, the deferent. The planet, P, is located on the epicycle, and the center of the deferent coincides with the center of the earth.

The epicycle-deferent system is intended to explain only motion with respect to the sphere of the stars. Both the epicycle and the deferent in Figure 19*a* are drawn on the plane of the ecliptic, so that the rotation of the stellar sphere carries the entire diagram (except the central earth) through one rotation per day and thus produces the diurnal motion of the planet. If the epicycle and deferent of the figure were stationary and did not have an additional motion of their own, the planet would be fixed in the plane of the ecliptic and would therefore have the motion of a zodiacal star, a westward circle executed once in every 23 hours 56 minutes. From now on, whenever reference is made to the motion of the deferent or the epicycle, it is the *additional* motion of these circles in the plane of the ecliptic that is meant. The diurnal rotation of the sphere and of the plane of the ecliptic will be taken for granted.

Suppose, for example, that the deferent rotates eastward once in a year and that the sun is placed on the deferent at the position now occupied by the center of the epicycle, the epicycle itself being removed. Then the rotation of the deferent carries the sun through its annual journey around the ecliptic, and the sun's motion has been analyzed, at least approximately, in terms of the motion of a single deferent in the plane of the ecliptic. This is the technique taken for granted in the explanation of average planetary motions in Figure 16.

Now imagine that the sun is removed and the epicycle is returned to its position on the deferent. If the epicycle rotates just three times around its moving center while the deferent rotates once and if the two circles rotate in the same direction, then the total motion of the planet within the sphere of the stars produced by the combined motions of the epicycle and the deferent is just the looped curve shown in Figure 19*b*. When the rotation of the epicycle carries the planet outside of the deferent, the motions of both the epicycle and the deferent combine to move the planet to the east. But when the motion of the epicycle places the planet well inside the deferent, the epicycle carries the planet westward, in opposition to the motion of the deferent. Therefore, when the planet is closest to the earth, the two motions may combine to produce a net westward or retrograde motion. In Figure 19*b* the planet retrogresses whenever it is on the interior part of one of the small loops; everywhere else the planet moves normally toward the east, but at a variable rate.

Figure 19c shows the motion of the planet through one of the loops as viewed against the sphere of the stars by an observer on earth. Since both observer and loop are in the same plane, that of the ecliptic, the observer cannot see the open loop itself. What he sees is merely the position of the planet against the background provided by the ecliptic. Thus as the planet moves from position 1 to 2 in Figures 19b and 19c, the observer sees it move along the ecliptic toward the east. As the planet approaches position 2, it appears to move more slowly, stopping momentarily at 2 and then moving westward along the ecliptic as it travels from 2 towards 3. Finally the westward journey of

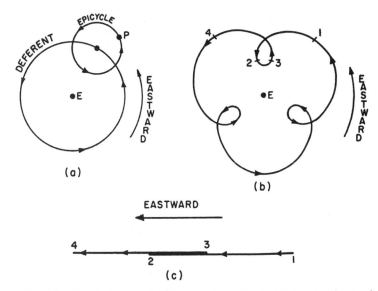

Figure 19. The basic epicycle-deferent system. A typical deferent and epicycle are shown in (a); the looped motion that they generate in the plane of the ecliptic is illustrated in (b); the third diagram (c) shows a portion (1–2–3–4) of the motion in (b) as it is seen by an observer on the central earth, E.

the planet on the ecliptic halts, and the planet again moves eastward, leaving position 3 on the loop and moving toward 4.

A system of one epicycle and one deferent therefore carries a planet around the ecliptic in an interval that, on the average, exactly equals the time required for one revolution of the deferent. The eastward motion is, however, interrupted and the planet temporarily moves westward at regular intervals equal to the time required for one revolution of the epicycle. The rates of revolution of the epicycle and deferent may be adjusted to fit the observations for any planet, yielding just that intermittent eastward motion among the stars which planets are observed to have. Furthermore, the epicycle-deferent system reproduces one other important qualitative feature of the appearances: a planet can retrogress only when its motion brings it nearest to the earth and that is the position in which the planet should and does appear brightest. Its great simplicity plus this novel explanation of varying planetary brilliance are the primary causes for the new system's victory over the older system of homocentric spheres.

The epicycle-deferent system described by Figure 19 incorporates one special simplification that is not characteristic of the motion of any planet. The epicycle is made to revolve *exactly* three times for each revolution of the deferent. Therefore, whenever the deferent completes one revolution, the epicycle returns the planet to the same position it occupied at the beginning of the revolution; the retrograde loops always occur at the same places; and the planet always requires the same amount of time to complete its trip around the ecliptic. When designed to fit the observations of real planets, however, epicycle-deferent systems never perform in quite this manner. For example, Mercury is observed to require an average of 1 year to complete a journey around the ecliptic, and it retrogresses once every 116 days. Therefore Mercury's epicycle must revolve just over three times while the deferent turns once; the epicycle completes three revolutions in 348 days which is less than the year required for a rotation of the deferent.

Figure 20a shows the path of a planet carried through one trip around the ecliptic by an epicycle that turns slightly more than three times for each rotation of its deferent. The planet starts in the middle of a retrograde loop and completes its third full loop before the deferent completes its first full rotation; the planet therefore averages slightly more than three retrograde loops in each trip around the ecliptic. If the motion of Figure 20a were continued through a second trip, the new set of retrograde loops would fall slightly to the west of those generated during the first trip. Retrograde motion would not occur at the same position in the zodiac on successive trips, and this is characteristic of the observed progress of planets along the ecliptic.

Figure 20b indicates a second characteristic of the motion generated by an epicycle that does not revolve an integral number of times in each revolution of the deferent. The planet at P in the figure is at the position closest to the earth, the position from which the journey of Figure 20a began. After one revolution of the deferent, the epicycle will have turned slightly more than three times, and the planet will have arrived at position P', so that it now appears to the west of its

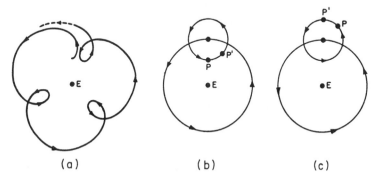

(a)   (b)   (c)

Figure 20. Motion generated by an epicycle and deferent when the epicycle turns slightly more than three times for each revolution of the deferent. The planet's path during a single complete journey through the stars is shown in (a). This journey requires more than one revolution of the deferent, as indicated by (b), which shows the planet's position at the beginning (P) and the end (P') of the deferent's first full revolution. Diagram (c) shows the planet's position at the beginning and end of a later revolution of the deferent, one that carries the planet more than once around the ecliptic.

starting point. The deferent must turn eastward through more than a single revolution to carry the planet fully around the ecliptic; the corresponding trip through the constellations therefore requires more time than the average. Others, however, require less. After several more revolutions of the deferent, each ending with the planet still farther from the earth, the planet might start a new journey from the new position P in Figure 20c. One more revolution of the deferent would carry the planet to P', a point to the east of P Since this revolution of the deferent carries the planet through more than one trip around the ecliptic, this journey is a particularly rapid one. Figures 20b and 20c represent very nearly the extreme values of the time required for a journey around the ecliptic; intermediate trips consume intermediate amounts of time; on the average, a journey around the ecliptic requires the same time as a rotation of the deferent. But the epicycle-deferent system allows for deviations from one trip to the next. Once again it provides an economical explanation of an observed irregularity of the planetary motions.

To describe the motions of all the planets a separate epicycle-deferent system must be designed for each. The motion of the sun and moon can be treated approximately by a deferent alone, for these planets do not retrogress. The sun's deferent turns once a year; the moon's revolves once in 27⅓ days. The epicycle-deferent system for Mercury is much like the one discussed above; the deferent turns once a year and the epicycle once in 116 days. By utilizing the observations recorded early in this chapter, we could design similar systems for other planets. Most of these would yield looped planetary paths like the one shown in Figure 20a. If the epicycle is larger relative to the deferent, the size of the loops is increased. If the epicycle turns more quickly relative to the speed of the deferent, then there are more loops included in one journey around the ecliptic. There are approximately eleven loops in each trip made by Jupiter, and approximately twenty-eight in each by Saturn. In short, by appropriate variations in the relative sizes and the speeds of the epicycle and the deferent, this system of compounded circular motions can be adjusted to fit approximately an immense variety of planetary motions. A properly designed combination of circles will even give a good qualitative account of the immense irregularities in the motion of an atypical planet like Venus (Figure 21).

Ptolemaic Astronomy

The discussion of the previous section illustrates the power and versatility of the epicycle-deferent system as a method for ordering and predicting the motions of the planets. But this is only the first step. Once the system was available to account for the most striking irregularities of planetary motion — retrogression and the irregular amounts of time consumed in successive journeys around the ecliptic — it became clear that there were still other, though very much smaller, irregularities to be considered.

Just as the two-sphere model provided a precise mechanism for the diurnal motions, thus permitting detailed study of the principal planetary irregularities, so the epicycle-deferent system, by providing an account of the main planetary motions, permitted the observational isolation of smaller irregularities. This is the first example of the con-

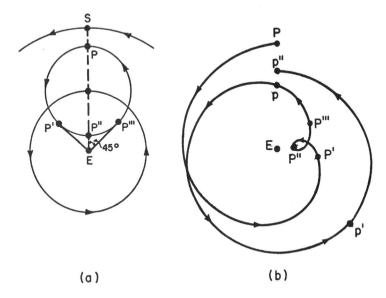

(a)   (b)

Figure 21. (a) A one-epicycle one-deferent system for Venus and (b) the motion that it generates in the plane of the ecliptic.

In (a), notice the following characteristics of the design: The deferent rotates once in a year, so that if the center of the epicycle is once aligned with the earth, E, and the center of the sun, S, it will stay in alignment forever, and Venus will never appear very far from the sun. The angles SEP' and SEP''' are the largest angles that can appear between the sun and Venus, and the condition that these angles of maximum elongation be 45° completely determines the relative sizes of the epicycle and deferent. The epicycle rotates once in 584 days, so that if Venus starts at P, close to the sun, it will arrive at P' (maximum elongation as an evening star) after 219 days (3/8 revolution); at P'' after 292 days (1/2 revolution); and at P''' (maximum elongation as a morning star) after 365 days (5/8 revolution).

The second diagram shows the path along which Venus is carried by the moving circles sketched in (a). Here P is the starting point, as in the first diagram; P' is Venus's position at maximum eastward elongation (219 days); P'' is the planet's location midway through a retrograde loop (292 days); and P''' is its position at maximum westward elongation (365 days). Venus's first journey around the ecliptic ends at p after 406 days (note the great length) and includes one retrogression and two maximum elongations. Its next trip (p to p' to p'') requires only 295 days and includes none of these characteristic phenomena. At p' Venus is again closest to the sun, a position reached after one complete revolution of the epicycle (584 days). This is, at least qualitatively, the way Venus does behave!

cept's fruitfulness. When the motion predicted by a one-epicycle one-deferent system is compared with the observed motion of an individual planet, it turns out that the planet is not always seen at quite the position on the ecliptic where the geometry of the model says it should be. Venus does not, if observed precisely, always attain its maximum deviation of 45° from the sun; the intervals between successive retrogressions of a single planet are not always quite the same; and none of the planets, except the sun, stays on the ecliptic throughout its motion. The one-epicycle one-deferent system was not, therefore, the final answer to the problem of the planets. It was only a very promising start and one that lent itself to both immediate and long-continued development. During the seventeen centuries that separate Hipparchus from Copernicus all the most creative practitioners of technical astronomy endeavored to invent some new set of minor geometric modifications that would make the basic one-epicycle one-deferent technique precisely fit the observed motion of the planets.

In antiquity the greatest of these attempts was made around A.D. 150 by the astronomer Ptolemy. Because his work displaced that of his predecessors and because all of his successors, including Copernicus, modelled their work upon his, the whole series of attempts for which Ptolemy provides the archetype is now usually known as Ptolemaic astronomy. That phrase, "Ptolemaic astronomy," refers to a traditional approach to the problem of the planets rather than to any one of the particular putative solutions suggested by Ptolemy himself, his predecessors, or his successors. Each of the particular individual solutions, and especially Ptolemy's, has an intense interest, both technical and historical, but both the particular solutions and their historical interrelationships are too complex to be considered here. Instead of attempting a general developmental account of the various Ptolemaic planetary systems, we shall therefore simply survey the main sorts of modification to which the basic epicycle-deferent system was subjected at various times between its first invention three centuries before Christ and its rejection by the followers of Copernicus.

Though their most important application is to the complex motions of the planets, the principal ancient and medieval modifications of the epicycle-deferent system are most simply described in their occasional applications to the apparently simpler motions of the sun and moon. The sun, for example, does not retrogress, so its motion does not require a major epicycle of the sort described in the last section. But fixing the sun on a deferent that rotates uniformly about the earth as center does not give a quantitatively precise account of the solar motion, for, as shown by a reëxamination of the dates of the solstices and equinoxes listed in Chapter 1, the sun takes almost 6 days longer to move from the vernal equinox to the autumnal equinox (180° along the ecliptic) than it does to move back from the autumnal equinox to the vernal equinox (again 180°). The sun's motion along the ecliptic is slightly more rapid during the winter than summer, and such a motion cannot be produced by a fixed point on a uniformly rotating earth-centered circle. Examine Figure 22a, in which the earth is shown at the center of a uniformly rotating deferent circle and in which the positions of the vernal and autumnal equinoxes on the sphere of the stars are indicated by the dashes VE and AE. Uniform rotation of the deferent will carry the sun, S, from VE to AE in the same time that it takes to carry it back from AE to VE, and this corresponds only approximately with observation.

Suppose, however, that the sun is removed from the deferent and placed on a small epicycle that rotates once westward while the deferent rotates once eastward. Eight positions of the sun in such a system are shown in Figure 22b. It is clear that the summer half of the deferent's rotation does not carry the sun the entire distance from VE to AE and that the winter half of the rotation carries the sun farther than the distance from AE to VE. So the effect of the epicycle is to increase the time spent by the sun in the 180° between VE and AE and to decrease

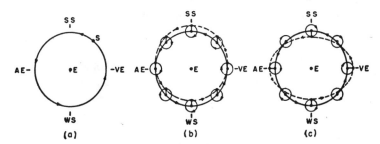

Figure 22. Functions of a minor epicycle. In (a) the sun, moved by a single earth-centered deferent, requires the same time to move from AE to VE that it needs to move back. In (b) the joint motion of deferent and minor epicycle carries the sun along the broken curve, so that more time is required for the trip from VE to AE than for the return. Diagram (c) shows the curve generated when the minor epicycle revolves at twice the rate used in constructing (b).

the time spent in the other half of the ecliptic between *AE* and *VE*. If the radius of the little epicycle is 0.03 the radius of the deferent, the difference in the time spent by the sun in the winter and the summer halves of the ecliptic will be the required 6 days.

The epicycle employed in the preceding discussion to correct a minor irregularity of the sun's motion is relatively small, and it produces no retrograde loops. Its function is therefore quite different from that of the larger epicycles considered in the last section, and, though Ptolemaic astronomers never did so, it will prove convenient to keep these two functions apart. Henceforth we shall use the term "major epicycle" for the large epicycles used to produce the qualitative appearance of retrograde motion and the term "minor epicycle" for the additional circles used to eliminate small quantitative discrepancies between theory and observation. All versions of the Ptolemaic system, both before and after Ptolemy, had just five major epicycles, and it is these with which Copernicus' reform did away. In contrast, the number of minor epicycles and similar devices needed to account for small quantitative discrepancies depended only on the precision of the available observations and on the accuracy of the predictions demanded from the system. The number of minor epicycles employed in the various versions of Ptolemaic astronomy therefore varied greatly from one version to the next. Systems employing half a dozen to a dozen minor epicycles were not uncommon in antiquity and the Renaissance, for by an appropriate choice of the size and speed of minor epicycles almost any sort of small irregularity could be explained away. That is why, as we shall see, Copernicus' astronomical system was so nearly as complex as Ptolemy's. Though his reform eliminated major epicycles, Copernicus was as dependent upon minor epicycles as his predecessors.

One sort of irregularity was treated with the aid of a minor epicycle in Figure 22*b*; another sort is shown in Figure 22*c*. There the minor epicycle rotates twice westward while the deferent moves once eastward. Combining the two rotations results in a total motion (broken line in the figure) along a flattened circle. A planet moving on this curve moves faster and spends less time in the vicinity of the summer and winter solstices than it does near the two equinoxes. If the epicycle had turned slightly less than twice while the deferent rotated once, then the positions on the ecliptic at which the planet's apparent speed was greatest would have changed on successive trips around the ecliptic. If it had appeared fastest near the summer solstice on one trip, it would have passed the summer solstice before gaining its greatest speed on the next trip. Other variations of this sort can be produced at will.

Uses of the minor epicycle are not limited to the nonretrogressing planets, the sun and moon. A minor epicycle can be placed upon a major epicycle and used in the prediction of the more elaborate planetary motions; in fact, planetary motions provided the minor epicycle's main astronomical application. One such application, an epicycle on an epicycle on a deferent, is shown in Figure 23a. If the major epicycle turns eight times eastward and the minor epicycle once westward during one rotation of the deferent, then the path within the sphere of the stars described by the planet is that shown in Figure 23b. It has eight normal retrograde loops, but these are somewhat more densely clustered in the half of the ecliptic between the vernal equinox and the autumnal equinox than in the half between the autumnal and vernal equinox. If the rate of rotation of the minor epicycle is now

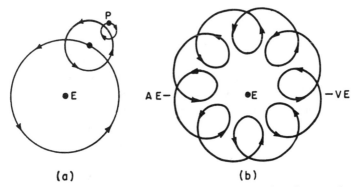

(a)   (b)

Figure 23. An epicycle on an epicycle on a deferent (a) and a typical path through space (b) generated by this system of compounded circles. For simplicity, the path has been shown rejoining itself smoothly, a situation that does not occur in the motion of real planets.

doubled, the path described by the planet is flattened as in Figure 22c. These diagrams begin to suggest the complexities of the paths that minor epicycles can produce.

Nor is a minor epicycle the only device available for correcting minor discrepancies between one-epicycle one-deferent systems and the observed behavior of the planets. A glance at Figure 22*b* indicates that the effect there produced by a minor epicycle that rotates westward once as the deferent turns through a single eastward rotation can equally well be achieved by a single deferent whose center is displaced from the center of the earth. Such a displaced circle, known to ancient astronomers as an eccentric, is shown in Figure 24*a*. If the distance

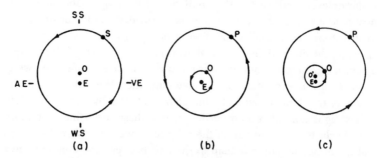

Figure 24. An eccentric (*a*), an eccentric on a deferent (*b*), and an eccentric on an eccentric (*c*).

between the earth, *E*, and the center of the eccentric, *O*, is about 0.03 the radius of the eccentric, this displaced circle will account for the 6 extra days that the sun spends between the vernal and autumnal equinoxes. That is the particular device that Ptolemy used in his own account of the sun. Other values of the distance *EO*, employed in conjunction with one or more epicycles, will account for other minor planetary irregularities. Additional effects may be obtained by placing the center of the eccentric on a small deferent (Figure 24*b*), or on a second smaller eccentric (Figure 24*c*). These two devices can be shown to be geometrically fully equivalent to a minor epicycle on a deferent and a minor epicycle on an eccentric, respectively, and most Ptolemaic astronomers used these small central circles in preference to minor epicycles. In all cases one or more epicycles may be added, and any or all of these circles may be tilted into different planes to account for the north and south deviations of the planets from the ecliptic.

One more device, the equant, was developed in antiquity to aid in the reconciliation of the theory of epicycles with the results of accurate observation. This device is of particular importance because Copernicus' aesthetic objections to it (Chapter 5) provided one essential motive for his rejection of the Ptolemaic system and his search for a radically new method of computation. Copernicus used epicycles and eccentrics like those employed by his ancient predecessors, but he did not use equants, and he felt that their absence from his system was one of its greatest advantages and one of the most forceful arguments for its truth.

One form of equant, designed, for simplicity of illustration, to account for the previously discussed irregularity in the sun's motion, is shown in Figure 25. The center of the sun's deferent coincides as before with the center of the earth, E, but the deferent's rate of rotation is now required to be uniform not with respect to its geometric center E, but with respect to an equant point, A, displaced in this case toward the summer solstice. That is, the angle a subtended at the equant point A by the sun and the summer solstice is required to change at a constant rate. If the angle increases by 30° in one month, then it must increase by 30° in every month of the same length. In the figure the sun is shown at the vernal equinox, VE. To reach the autumnal equinox, AE, it must complete a semicircle, which will change the angle a by more than 180°, and to return from AE to VE it must complete a second semicircle, which will change a by less than 180°. Since every 180° increase in a requires the same amount of time, the sun must take longer to go from VE to AE than it requires for the return journey from AE to VE. Therefore, viewed from the equant point A, the sun travels at an irregular rate, fastest near the winter solstice and slowest near the summer solstice.

That is the defining feature of the equant. The rate of rotation of a deferent or some other planetary circle is required to be uniform, not with respect to its own geometric center, but with respect to an equant point displaced from that center. Observed from the geometric center of its deferent, the planet seems to move at an irregular rate or to wobble. Because of the wobble, Copernicus felt that the equant was not a legitimate device for application to astronomy. For him the apparent irregularities of the rotation were violations of the uniform circular symmetry that made the system of epicycles, deferents, and eccentrics so plausible and attractive. Since the equant was normally applied to eccentrics and since similar devices occasionally made the epicycle wobble as well, it is not hard to imagine how Copernicus might have considered this aspect of Ptolemaic astronomy monstrous.

The mathematical devices sketched in the preceding pages were not all developed at a stroke or by Ptolemy. Apollonius, in the third century B.C., knew both major epicycles (Figure 19a) and eccentrics with moving centers (Figure 24b). During the following century Hipparchus added minor epicycles and a more general theory of eccentrics to the arsenal of astronomical weapons. In addition he combined these devices to provide the first quantitatively adequate account of the irregularities in the motions of the sun and moon. Ptolemy himself added the equant, and during the thirteen centuries between his time and that of Copernicus, first Moslem and then European astronomers employed still other combinations of circles — including the epicycle on an epicycle (Figure 23a) and the eccentric on an eccentric (Figure 24c) — to account for additional planetary irregularities.

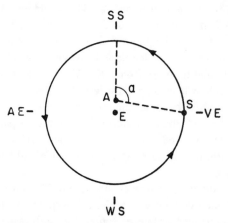

Figure 25. The equant. The sun, S, moves on the earth-centered circle but at an irregular rate determined by the condition that the angle a vary uniformly with time.

But Ptolemy's contribution is the outstanding one, and this entire technique of resolving the problem of the planets is appropriately known by his name, because it was Ptolemy who first put together a particular set of compounded circles to account, not merely for the motions of the sun and moon, but for the observed quantitative regularities and irregularities in the apparent motions of all the seven planets. His *Almagest*, the book that epitomizes the greatest achievements of ancient astronomy, was the first systematic mathematical treatise to give a *complete, detailed,* and *quantitative* account of all the celestial motions. Its results were so good and its methods so powerful that after Ptolemy's death the problem of the planets took a new form. To increase the accuracy or simplicity of planetary theory Ptolemy's successors added epicycles to epicycles and eccentrics to eccentrics, exploiting all the immense versatility of the fundamental Ptolemaic technique. But they seldom or never sought fundamental modifications of that technique. The problem of the planets had become simply a problem of design, a problem to be attacked principally by the re-arrangement of existing elements. What particular combination of de-ferents, eccentrics, equants, and epicycles would account for the planetary motions with the greatest simplicity and precision?

We cannot pursue further the individual quantitative solutions of this problem proposed by Hipparchus, Ptolemy, and their successors. The complete quantitative systems are mathematically too complex. Much of Ptolemy's *Almagest* consists of quantitative mathematical tables, diagrams, formulas, and proofs, of long illustrative computations, and of lists of numerous observations. Yet the problems that set Coper-nicus to searching for a new approach to the problem of the planets and the advantages that he claimed to derive from his new system all lie within this abstruse body of quantitative theory. Copernicus did not attack the two-sphere universe, though his work ultimately overthrew it, and he did not abandon the use of epicycles and eccentrics, though these too were abandoned by his successors. What Copernicus did attack and what started the revolution in astronomy was certain of the ap-parently trivial mathematical details, like equants, embodied in the complex mathematical systems of Ptolemy and his successors. The initial battle between Copernicus and the astronomers of antiquity was fought over technical minutiae like those sketched in this section.

The Anatomy of Scientific Belief

For its subtlety, flexibility, complexity, and power the epicycle-deferent technique sketched in the two preceding sections has no parallel in the history of science until quite recent times. In its most developed form the system of compounded circles was an as-tounding achievement. *But it never* quite *worked.* Apollonius' initial conception solved the primary planetary irregularities — retrograde motion, variation of brightness, alteration in the time required for successive journeys around the ecliptic — and it did so simply and at a stroke. But it also disclosed some residual secondary irregularities. Some of these were explained away by the more elaborate system of compounded circles developed by Hipparchus, but still the theory did not quite match the results of observation. Even Ptolemy's complex combination of deferents, eccentrics, epicycles, and equants did not precisely reconcile theory and observation, and Ptolemy's was neither the most complex nor the last version of the system. Ptolemy's many successors, first in the Moslem world and then in medieval Europe, took up the problem where he had left it and sought in vain for the solution that had evaded him. Copernicus was still grappling with the same problem.

There are many variations of the Ptolemaic system besides the one that Ptolemy himself embodied in the *Almagest*, and some of them achieved considerable accuracy in predicting planetary positions. But the accuracy was invariably achieved at the price of complexity — the addition of new minor epicycles or equivalent devices — and increased complexity gave only a better approximation to planetary motion, not finality. No version of the system ever quite withstood the test of additional refined observations, and this failure, combined with the total disappearance of the conceptual economy that had made cruder versions of the two-sphere universe so convincing, ultimately led to the Copernican Revolution.

But the Revolution was an incredibly long time coming. For almost 1800 years, from the time of Apollonius and Hipparchus until the birth of Copernicus, the conception of compounded circular orbits within an earth-centered universe dominated every technically developed attack upon the problem of the planets, and there were a great many such attacks before Copernicus'. Despite its slight but recognized inaccuracy and its striking lack of economy (contrast the earlier two-sphere universe described in Chapter 1), the developed Ptolemaic system had an immense life span, and the longevity of this magnificent but clearly imperfect system poses a pair of closely related puzzles: How did the two-sphere universe and the associated epicycle-deferent planetary theory gain so tight a grip upon the imagination of the astronomers? And, once gained, how was the psychological grip of this traditional approach to a traditional problem released? Or to put the same questions more directly: Why was the Copernican Revolution so delayed? And how did it come to pass at all?

These are questions about the history of a particular set of ideas, and as history they will be considered at some length below. But they are also more generally concerned with the nature and structure of conceptual schemes and with the process by which one conceptual scheme replaces another. It is therefore illuminating to approach them first by returning briefly to the abstract logical and psychological categories introduced in the penultimate section of the first chapter. We there examined the functions of a conceptual scheme: we now ask how a smoothly functioning scheme, like the early two-sphere universe, can be replaced. Examine the logic of the phenomenon first.

Logically there are always many alternative conceptual schemes capable of bringing order to any *prescribed* list of observations, but these alternatives differ in their predictions about phenomena not included on the list. Both the Copernican and the Newtonian systems will account for naked-eye stellar and solar observations just as adequately as will the two-sphere system; Heraclides' system will do the same and so will the system developed by Copernicus' successor, Tycho Brahe; in theory there are an infinite number of other alternatives besides. But these alternatives agree principally about observations that have already been made. They do not give identical accounts of all possible observations. The Copernican system, for example, differs from the two-sphere universe in predicting an apparent annual motion of the stars, in demanding a much larger diameter for the stellar sphere, and in suggesting (though not to Copernicus) a new sort of solution for the problem of the planets. It is because of differences like these (and there are many others besides) that a scientist must believe in his system before he will trust it as a guide to fruitful investigations of the *unknown*. Only one of the different alternatives can *conceivably* represent reality, and the scientist exploring new territory must feel confident that he has chosen that one or the closest of the available approximations to it. But the scientist pays a price for this commitment to a particular alternative: he may make mistakes. A single observation incompatible with his theory demonstrates that he has been employing the wrong theory all along. His conceptual scheme must then be abandoned and replaced.

That, in outline, is the logical structure of a scientific revolution. A conceptual scheme, believed because it is economical, fruitful, and cosmologically satisfying, finally leads to results that are incompatible with observation; belief must then be surrendered and a new theory adopted; after this the process starts again. It is a useful outline, because the incompatibility of theory and observation is the ultimate source of every revolution in the sciences. But historically the process of revolution is never, and could not possibly be, so simple as the logical outline indicates. As we have already begun to discover, observation is never *absolutely* incompatible with a conceptual scheme.

To Copernicus the behavior of the planets was incompatible with the two-sphere universe; he felt that in adding more and more circles his predecessors had simply been patching and stretching the Ptolemaic system to force its conformity with observations; and he believed that the very necessity for such patching and stretching was clear evidence that a radically new approach was imperatively required. But Copernicus' predecessors, to whom exactly the same sorts of instruments and observations were available, had evaluated the same situation quite differently. What to Copernicus was stretching and patching was to them a natural process of adaptation and extension, much like the process which at an earlier date had been employed to incorporate the motion of the sun into a two-sphere universe designed initially for the earth and stars. Copernicus' predecessors had little doubt that the system would ultimately be made to work.

In short, though scientists undoubtedly do abandon a conceptual scheme when it seems in irreconcilable conflict with observation, the emphasis on logical incompatibility disguises an essential problem. What is it that transforms an apparently temporary discrepancy into an inescapable conflict? How can a conceptual scheme that one generation admiringly describes as subtle, flexible, and complex become for a later generation merely obscure, ambiguous, and cumbersome? Why do scientists hold to theories despite discrepancies, and, having held to them, why do they give them up? These are problems in the anatomy of scientific belief. They are the primary concern of the next two chapters, which set the stage for the Copernican Revolution proper.

Our immediate problem, however, is the analysis of the grip exerted upon men's minds by the ancient tradition of astronomical research. How could this tradition provide a set of mental grooves that guided the astronomical imagination, limited the conceptions available in research, and made certain sorts of innovations difficult to conceive and more difficult to accept? We have already dealt, at least implicitly, with the strictly astronomical aspects of this problem. Both the two-sphere universe and the associated epicycle-deferent technique were initially highly economical and fruitful; their first successes seemed to guarantee the fundamental soundness of the approach; surely only minor modifications would be required to make the mathematical predictions correspond with observation. A conviction of this sort is difficult to break, particularly once it has been embodied in the practice of a whole generation of astronomers who transmit it to their successors through their teaching and writing. This is the band-wagon effect in the realm of scientific ideas.

The band-wagon effect is not, however, the whole explanation of the strength of the astronomical tradition, and in trying to complete the explanation we shall be temporarily led away from astronomical problems altogether. The two-sphere universe provided a fruitful guide to the solution of problems outside as well as inside astronomy. By the end of the fourth century B.C. it had been applied not only to the problem of the planets, but also to terrestrial problems, like the fall of a leaf and the flight of an arrow, and to spiritual problems, like the relation of man to his gods. If the two-sphere universe, and particularly the conception of a central and stable earth, then seemed the indubitable starting point of all astronomical research, this was primarily because the astronomer could no longer upset the two-sphere universe without overturning physics and religion as well. Fundamental astronomical concepts had become strands in a far larger fabric of thought, and the nonastronomical strands could be as important as the astronomical in binding the imagination of astronomers. The story of the Copernican Revolution is not, therefore, simply a story of astronomers and the skies.

3

THE TWO-SPHERE UNIVERSE
IN ARISTOTELIAN THOUGHT

The Aristotelian Universe

In order to examine the ancient world view, in which astronomical and nonastronomical concepts were woven into a single coherent conceptual fabric, we must reverse chronological order and return for a while to the middle of the fourth century B.C. At that time the technically developed attack upon the problem of the planets had scarcely begun, but the same two-sphere cosmology that was guiding the mathematical researches of planetary astronomers had already acquired essential nonastronomical functions. Many of these can be discovered in the voluminous works of the great Greek philosopher and scientist, Aristotle (384–322 B.C.), whose immensely influential opinions later provided the starting point for most medieval and much Renaissance cosmological thought.

Aristotle's writings, which have reached us only in imperfect and highly edited form, deal with the scientific subjects now called physics, chemistry, astronomy, biology, and medicine, as well as with such nonscientific fields as logic, metaphysics, politics, rhetoric, and literary criticism. To each of these, particularly biology, logic, and metaphysics, he contributed new ideas that were uniquely his own, but even more important than his many scattered substantive contributions was his organization of all knowledge into a systematic and coherent whole. He was not quite successful; it is not difficult to find inconsistencies and occasional contradictions within the body of Aristotle's writings. But there is a fundamental unity in his view of man and the universe that has never since been achieved in a synthesis of comparable scope and originality. That is one of the reasons why his writings had such immense influence; others will be examined at the end of this chapter.

We first, however, require a brief structural sketch of the Aristotelian universe itself, followed by a more detailed discussion of the multiple functions attributed to the terrestrial and celestial spheres in Aristotelian thought.

For Aristotle the entire universe was contained within the sphere of the stars, or, more precisely, within the outer surface of the sphere. At every point inside the sphere there was some sort of matter — no holes or vacuums could exist in Aristotle's universe. Outside the sphere there was nothing — no matter, no space, nothing at all. In Aristotelian science matter and space go together; they are two aspects of the same phenomenon; the very notion of a vacuum is absurd. That is how Aristotle managed to explain the finite size and the uniqueness of the universe. Matter and space must end together: one need not build a wall to bound the universe and then wonder what bounds the wall. As Aristotle put it in his book, *On the Heavens*:

> It is plain, then, . . . that there is not, nor do the facts allow there to be, any bodily mass beyond the heaven. The world in its entirety is made up of the whole sum of available matter. . . , and we may conclude that there is not now a plurality of worlds, nor has there been, nor could there be. This world is one, solitary, and complete. It is clear in addition that there is neither place nor void . . . beyond the heaven; for in all place there is a possibility of the presence of body, [and] void is defined as that which, although at present not containing body, can contain it. . . .[1]

Like Plato's universe, part of which was briefly described in Chapter 1, the Aristotelian universe is self-contained and self-sufficient, leaving nothing outside itself. But Aristotle differentiates the interior of the universe in far more detail than Plato. The largest part of the interior is filled with a single element, the aether, which aggregates in a homocentric set of nesting shells to form a gigantic hollow sphere whose surfaces are the outside of the sphere of the stars and the inner surface of the homocentric sphere carrying the lowest planet, the moon. Aether is the celestial element — a crystalline solid in Aristotle's writings, though its solidity was frequently questioned by his successors. Unlike the substances known on earth, it is pure and unalterable, transparent and weightless. From it are made the planets and stars as well as the nest of concentric spherical shells whose rotations account for the celestial motions.

Between the times of Aristotle and Copernicus a number of different views prevailed about the form and the physical reality of these celestial spheres that moved the heavens. Aristotle's view was the most detailed and explicit of all. He believed that there were just fifty-five real crystalline shells made of aether and that these shells embodied in a physical mechanism the mathematical system of homocentric spheres developed by Eudoxus and his successor, Callippus. Aristotle almost doubled the number of spheres used by the earlier mathematicians, but the spheres that he added were mathematically superfluous. Their function was to provide the mechanical linkages necessary to keep the whole set of concentric shells turning; they transformed the entire nest of spheres into a gigantic piece of celestial clockwork, driven from the sphere of the stars. Since the universe was full, all spheres were in contact, and the rubbing of sphere on sphere provided drive for the entire system. The sphere of the stars drove its nearest interior neighbor, the outermost of the seven homocentric shells that moved Saturn. That shell drove its next interior neighbor in Saturn's set, and so on, until the motion was finally transmitted to the lowest sphere in the set that carried the moon. This was the innermost of the aetherial shells, the lower boundary of the celestial or superlunary region.

The set of epicycles and deferents, which replaced homocentric spheres for purposes of mathematical astronomy, did not fit very well into crystalline spheres like those proposed by Aristotle. As a result the attempt to find a *mechanical* explanation of the epicyclic motions was often neglected after the fourth century B.C., and the real existence of crystalline spheres was occasionally questioned. It is not, for example, clear from the *Almagest* whether Ptolemy believed in them at all. But in the period separating Ptolemy and Copernicus most educated people, including astronomers, seem to have believed in at least a bastard version of Aristotle's spheres. They allowed one spherical shell for the stars and one for each planet, and they supposed that each planetary shell was just thick enough for the planet to be at its inner surface when closest to the central earth and at its outer surface when farthest from the earth. These eight spheres were nested one inside the other to fill the entire celestial region. The motion of the sphere of the stars provided a precise explanation of the diurnal circles of the stars. The continual rotation of the seven planetary spheres explained the average motion, but only the average motion, of the planets.

Men who did not know or care about the irregularities of planetary motion could take the thick spheres quite literally: each planet was fixed in and carried around by its sphere. Planetary astronomers used epicycles, deferents, equants, and eccentrics to account for each planet's motion within its own thick spherical shell. For them the shells usually had at least metaphorical reality, but they rarely bothered with a physical explanation of a planet's motion within its sphere.

Five centuries or more after Aristotle's death this conception of thick nested spherical shells added one important technical ingredient to the astronomy of the post-Ptolemaic period. It enabled astronomers to compute the actual size of the individual planetary spheres and therefore of the universe as a whole. Observations of a planet's motion among the stars enables an astronomer to determine only the *relative sizes* of its epicycle and deferent or the *relative amount* of its eccentricity. Shrinking or expanding a planet's system of compounded circles does not change the position at which the planet appears against the ecliptic, provided that the relative dimensions of epicycle, deferent, and eccentric are held constant. But if each spherical shell must be just thick enough to contain the planet both when it is closest and when it is farthest from the earth, then a knowledge of the relative dimensions of epicycle, deferent, and so on, suffices to determine the ratio of the inside and outside diameters of each sphere. Furthermore, if the spheres must nest so that they fill the entire celestial region, the outer diameter of one must be equal to the inner diameter of the next, and the ratio of the distances to all the intershell boundaries can be computed. Finally, these relative dimensions can be transformed to absolute distances by utilizing the distance to the sphere of the moon determined during the second century B.C. by the technique discussed in the Technical Appendix, Section 4.

Estimates of size based on the conception of space-filling spheres just large enough to house each planet's set of epicycles and other circles do not appear in the astronomical literature until after Ptolemy's death, presumably because the first planetary astronomers were skeptical about the reality of such spheres. But after the fifth century A.D., estimates of this sort became quite common, and once again they helped to make the entire cosmology seem real. One widely known set of cosmological dimensions was provided by the Arab astronomer Al Fargani, who lived in the ninth century A.D. According to his calculation the outer surface of the moon's sphere was $64\frac{1}{6}$ earth's radii from the center of the universe, the outer surface of Mercury's sphere was 167 earth's radii from the center; of Venus's, 1120 earth's radii; of the sun's, 1220; of Mars's, 8867; of Jupiter's, 14405; and of Saturn's, 20110. Since Al Fargani gave the earth's radius as 3250 Roman miles, he placed the sphere of the stars more than 75 million miles from the earth. It is an immense distance but, by modern theory, almost a million times smaller than the distance to the nearest star.

A glance at Al Fargani's measurements shows that the terrestrial region, the space below the underside of the moon's sphere, is but a minuscule portion of the universe. Most of space is the heavens; most matter is the aether of the crystalline spheres. But the small size of the sublunary region does not make it unimportant. Even in Aristotle's version, and to a far greater extent in the medieval Christian revision of Aristotelian cosmology, the tiny central core of the universe is the kernel for which the rest was made. It is man's abode, and its character is very different from that of the celestial region above it.

The sublunary region is filled not with one element but with four (or, in later writers, some other small number), and the distribution of these four terrestrial elements, though simple in theory, is extremely complex in fact. According to the Aristotelian laws of motion, to be discussed below, the elements would, in the absence of any external pushes and pulls upon them, settle into a series of four concentric shells like the aetherial spheres of the fifth element surrounding them. Earth, the heavy element, would move naturally into a sphere at the geometric center of the universe. Water, also heavy but not so heavy as earth, would settle in a spherical shell about the central region of earth. Fire, the lightest element, would rise spontaneously to form a shell of its own immediately below the moon's sphere. And air, also a light element, would complete the structure by filling the remaining shell between water and fire. Having achieved these positions, the elements would rest in place with their full elemental purity. Left to itself, undisturbed by outside forces, the sublunary region would be a static region, mirroring the heavenly spheres in its structure.

But the terrestrial region is never undisturbed. It is bounded by the moving sphere of the moon, and the movement of this boundary constantly pushes the layer of fire beneath it, setting up currents which jostle and mix the elements together throughout the sublunary world. Therefore, the elements can never be observed in their pure form. The continual chain of pushes, deriving immediately from the sphere of the moon and ultimately from the sphere of the stars, keeps them mixed together in various and varying proportions. The structure of shells is still approximated; the appropriate element predominates in the appropriate region. But each element contains at least traces of the others, and these transform its character, giving rise, according to the proportions of the mixture, to all of the varied substances that can be discovered on the earth. The motions of the heavens are therefore responsible for all change and almost all variety observed in the sublunary world.

It is within this Aristotelian universe, whose scope and adequacy are scarcely represented by the preceding sketch, that we must search for the strength of the pre-Copernican astronomical tradition. Why, despite the real difficulties encountered by the Ptolemaic system, did astronomers continue for so long to assume that the earth had to be the stable center of the universe and of at least the average planetary orbits? One familiar answer to this question is already apparent: Aristotle, the greatest philosopher-scientist of antiquity, had declared the earth to be immobile, and his word was taken with immense seriousness by his successors, for many of whom he became "the Philosopher," the first authority on all questions of science and cosmology.

But Aristotle's authority, though important, is only the beginning of an answer, because Aristotle said a great many things which later philosophers and scientists did not have the least difficulty in rejecting. In the ancient world there were other schools of scientific and cosmological thought, apparently little influenced by Aristotelian opinion. Even in the late centuries of the Middle Ages, when Aristotle did become the dominant authority on scientific matters, learned men did not hesitate to make drastic changes in many isolated portions of his doctrine. The list of alterations introduced by later Aristotelians into Aristotle's original teachings is almost endless, and some of these alterations were far from trivial. As we shall see in the next chapter, a few of the criticisms directed at Aristotle by his successors play a direct and causal role in the Copernican Revolution.

Yet no later Aristotelian suggested that the earth was a planet or that it was located away from the center of the universe. That innovation proved a peculiarly difficult one for an Aristotelian to conceive or to accept, because the concept of a unique central earth was interwoven with so many other important concepts within the fabric of Aristotelian thought. An Aristotelian universe can be built as well with three or five terrestrial elements as with four and very nearly as well with epicycles as with homocentric spheres, but it cannot and did not survive the modification that made the earth a planet. Copernicus tried to design an essentially Aristotelian universe around a moving earth, but he failed. His followers saw the full consequences of his innovation, and the entire Aristotelian structure crumbled. The concept of a central and stable earth was one of the few major constitutive concepts in a closely knit and coherent world view.

The Aristotelian Laws of Motion

A first example of the integration of astronomical and non-astronomical thought is provided by Aristotle's explanation of terrestrial motion. As we have already noted, Aristotle believed that, in the absence of external pushes derived ultimately from the heavens, each of the terrestrial elements would remain at rest in that part of the terrestrial region natural to it. Earth rests naturally at the center, fire at the periphery, and so on. In fact, the elements and the bodies composed of them are constantly wrested from their natural positions. But that requires the application of a force; an element resists displacement; and, once displaced, it strives to regain its natural position by the shortest possible path. Pick up a rock or some other earthy material and feel it tug away, attempting to reach its natural position at the geometric center of the universe. Or watch the flames of a fire leap upward on a clear night as they strive for their natural place at the periphery of the terrestrial region.

We shall examine later the psychological sources and the strength of this Aristotelian explanation of terrestrial motion. But first notice the bulwark that these doctrines, drawn from terrestrial physics, provide for the earth-centered universe of the astronomer. In an important passage from *On the Heavens* Aristotle derives the sphericity, stability, and central location of the earth from them. We have previously seen them derived by astronomical arguments, but note how secondary a role astronomical considerations play here.

The natural motion of the earth as a whole, like that of its parts, is towards the center of the Universe: that is the reason why it is now lying at the center. It might be asked, since the center of both is the same point, in which capacity the natural motion of heavy bodies, or parts of the earth, is directed towards it; whether as center of the Universe or of the earth. But it must be towards the center of the Universe that they move. . . . It so happens that the earth and the Universe have the same center, for the heavy bodies do move also towards the center of the earth, yet only incidentally, because it has its center at the center of the Universe. . . .

From these considerations it is clear that the earth does not move, neither does it lie anywhere but at the center. In addition the reason for its immobility is clear from our discussions. If it is inherent in the nature of earth to move from all sides to the center (as observation shows), and of fire to move away from the center towards the extremity, it is impossible for any portion of earth to move from the center except under constraint. . . . If then any particular portion is incapable of moving from the center, it is clear that the earth itself as a whole is still more incapable, since it is natural for the whole to be in the place towards which the part has a natural motion. . . .

Its shape must be spherical. . . . To grasp what is meant we must imagine the earth as in process of generation. . . . It is plain, first, that if particles are moving from all sides alike towards one point, the center, the resulting mass must be similar on all sides; for if an equal quantity is added all round, the extremity must be at a constant distance from the center. Such a shape is a sphere. But it will make no difference to the argument even if the portions of the earth did not travel uniformly from all sides towards the center. A greater mass must always drive on a smaller mass in front of it, if the inclination of both is to go as far as the center, and the impulsion of the less heavy by the heavier persists to that point. . . .

Further proof is obtained from the evidence of the senses. (i) If the earth were not spherical, eclipses of the moon would not exhibit segments of the shape they do. . . . (ii) Observation of the stars also shows not only that the earth is spherical but that it is of no great size, since a small change of position on our part southward or northward visibly alters the circle of the horizon, so that the stars above our heads change their position considerably, and we do not see the same stars as we move to the North or South. Certain stars are seen in Egypt and the neighborhood of Cyprus, which are invisible in more northerly lands, and stars which are continuously visible in the northern countries are observed to set in the others. This proves both that the earth is spherical and that its periphery is not large, for otherwise such a small change of position could not have had such an immediate effect. For this reason those who imagine that the region around the Pillars of Heracles joins on to the regions of India, and that in this way the ocean is one, are not, it would seem, suggesting anything utterly incredible.[2]

Passages like this indicate that astronomy and terrestrial physics are not independent sciences. Observations and theories developed for one become intimately entangled with those drawn from another. Therefore, though difficulty in solving the problem of the planets might have provided an astronomer with a motive for experimenting *in astronomy* with the conception of a moving earth, he could not do so without upsetting the accepted basis of terrestrial physics in the process. The very notion of a moving earth would be unlikely to occur to him, because, for reasons drawn from his nonastronomical knowledge, the conception seemed so implausible. That seems to be what Ptolemy and his successors meant when they later described the astronomical hypotheses of Aristarchus, Heraclides, and the Pythagoreans as "ridiculous" even though astronomically satisfactory.

Examine, for example, the following passage from the *Almagest*, in which Ptolemy rejects Heraclides' theory that the sphere of the stars is stationary and that its apparent westward diurnal motion is due to a real eastward diurnal rotation of the central earth. Ptolemy begins with arguments for the sphericity and central position of the earth much like those given by Aristotle in the passage quoted above. Then he continues:

Certain thinkers, though they have nothing to oppose to the above arguments, have concocted a scheme which they consider more acceptable, and they think that no evidence can be brought against them if they suggest for the sake of argument that the heaven is motionless, but that the earth rotates about one and the same axis from west to east, completing one revolution approximately every day. . . .

These persons forget however that, while, so far as appearances in the stellar world are concerned, there might, perhaps, be no objection to this theory. . . , yet, to judge by the [terrestrial] conditions affecting ourselves and those in the air about us, such a hypothesis must be seen to be quite ridiculous. . . . [If the earth] made in such a short time such a colossal turn back to the same position again, . . . everything not actually standing on the earth must have seemed to make one and the same movement always in the contrary sense to the earth, and clouds and any of the things that fly or can be thrown could never be seen travelling towards the east, because the earth would always be anticipating them all and forestalling their motion towards the east, insomuch that everything else would seem to recede towards the west and the parts which the earth would be leaving behind it.[8]

The gist of Ptolemy's argument is the same as Aristotle's, and many other arguments were derived from the same principles during the Middle Ages and the Renaissance. Unless it is pushed, a body will move straight toward its natural position and then rest there. These natural positions and the lines by which bodies move to them are determined entirely by the intrinsic geometry of an absolute space, a space in which each position and direction is permanently labeled whether or not the position is occupied. Therefore, as Aristotle says elsewhere in *On the Heavens*, "If the earth were removed to where the moon is now, separate parts of it would not move towards the whole, but towards the place [the center] where the whole is now."[4] The natural motion of a stone is governed by space alone, not by the stone's relation to other bodies. Therefore, a stone thrown vertically upward moves away and returns along a straight line fixed once and for all in space, and if the earth moves while the stone is in the air, the stone will not rejoin the earth at the point from which it departed. By the same token, clouds which already occupy their natural positions would be left behind as the earth rotates. Only if the moving earth carries the air with it, could a cloud or stone follow the earth at all, and even the motion of the air would not push a stone hard enough to keep it in step with the earth's rotation.

There are, of course, difficulties in this Aristotelian theory of motion, and some of them will later play a significant role in the Copernican Revolution. But, like the two-sphere universe itself, Aristotle's theory of motion is an excellent first step toward an understanding of motion, and it does necessitate a central stationary earth. Advocates of a planetary earth will therefore require a new theory of motion, and until such a theory is invented, as it was during the Middle Ages, a knowledge of terrestrial physics will inhibit the astronomical imagination.

The Aristotelian Plenum

A second illustration of the blinders fitted to the astronomer by the coherent interrelations between his astronomical and nonastronomical knowledge is provided by the Aristotelian conception of a full universe or plenum. This example is more typical than the last, for the ties between the various strands of knowledge are here both more numerous and less binding than those illustrated above. The complex pattern of Aristotelian thought now begins to emerge.

The ancient conception of the fullness of the universe is often referred to as the *horror vacui*, nature's abhorrence of a vacuum. As an
explanatory principle it can be paraphrased to read: Nature will always act to prevent the formation of a vacuum. In this form the
Greeks derived it from and used it to explain a large variety of natural
phenomena. Water will not flow from an open bottle with a small neck
unless a second hole is made in the bottle, because without a second
hole at which air can enter the emerging water would leave a vacuum
behind it. Siphons, water clocks, and pumps were economically explained on the same basis. Some ancient thinkers applied the *horror
vacui* to the explanation of adhesion and to the design of hot-air and
steam engines. The experimental basis of the principle could not be
challenged. Convincing approximations to vacuums cannot be produced on earth without apparatus of which the Greeks knew nothing.
There were no pneumatic phenomena to challenge the principle until,
with the large-scale development of deep mining during the sixteenth
century, it was discovered that lift pumps will not raise water more
than 30 feet. Rejecting the *horror vacui* necessarily meant destroying
a thoroughly satisfactory scientific explanation of a host of terrestrial
phenomena.

For Aristotle and most of his successors, however, the *horror vacui*
was more than a successful experimental principle applicable on and
near the surface of the earth. Aristotle held not only that there are *in
fact* no vacuums in the terrestrial world, but that there can *in principle* be no vacuums anywhere in the universe. The very concept of
a vacuum was to him a contradiction in terms, like the concept "square
circle." Today, when everybody has seen a "vacuum" tube and heard
of a "vacuum" pump, Aristotle's logical proofs of the impossibility of
a void convince almost no one, though it is frequently difficult to discover the faults in his arguments. But in the absence of the experimental counterevidence, which we now possess, they seemed convincing, for they arose from a genuine difficulty inherent in the words
with which we discuss problems of matter and space. Apparently space
can be defined only in terms of the volume occupied by body. In the
absence of material body there is nothing in terms of which to define
space; it cannot apparently exist by itself at all. Matter and space
are inseparable, two sides of the same coin. There can be no space
without matter. In Aristotle's more cumbersome words, "there is no

such thing as a dimensional entity, other than that of material substances." [5]

The theory of a full universe therefore entered ancient science with the combined authority of logic and experiment, and it immediately became an essential ingredient of cosmological and astronomical theories. It is, for example, involved in the Aristotelian explanation of the endurance of motion within the sphere of the stars. If any one of the celestial or terrestrial shells were replaced by a void, all motion within that shell would cease. The rubbing of shell on shell produces all motion, except return to natural position, and a void would break the chain of pushes. Again, as we have already noted, the impossibility of a void is the basis of the universe's finitude. Beyond the sphere of the stars is neither space nor matter — nothing at all. Without a concept that indissolubly united matter and space, the Aristotelian would be forced to admit the infinity of the universe. Matter could be bounded by void, and void could, in turn, be bounded by matter, but there could be no terminus, no last boundary at which the universe ended once and for all.

But an infinite universe could scarcely remain an Aristotelian universe for two reasons. An infinite space has no center: every point is equally distant from all points on the periphery. And if there is no center, there is no preferred point at which the heavy element earth can aggregate, and there is no intrinsic "up" and "down" to determine the natural motion of an element returning to its proper place. In fact there is no "natural place" in an infinite universe, for each place is like every other. The whole Aristotelian theory of motion is, as we shall see more fully later, inextricably bound to the conception of a finite and fully occupied space. The two stand or fall together.

Nor are these the only difficulties presented to an Aristotelian by the infinitude of space. If space is infinite and there is no special center point, it is scarcely plausible that all the earth, water, air, and fire in the universe should have aggregated at one and only one point. In an infinite universe, it is natural to presume that there are other worlds scattered here and there through all of space. Perhaps there are plants, men, and animals on these other worlds. Thus the earth's uniqueness vanishes; the peripheral force that drives the whole disappears with it; man and the earth cease to be at the focus of the universe. Both in antiquity and in the Middle Ages, most of those philosophers who,

like the atomists, believed that the universe was infinite felt themselves impelled to accept the reality of the void and a plurality of
worlds as well. And until the seventeenth century no one who embraced
this set of concepts produced a cosmology able to compete with the
Aristotelian in the explanation of everyday terrestrial and celestial
phenomena. The infinite universe may be a common-sense universe
today, but today common sense has been reëducated.

The multifarious roles of the conception of a full universe in Aristotelian thought is our one full-dress example of the coherence of a
cosmology or a world view. The plenum is implicated in pneumatics,
the endurance of motion, the finitude of space, the laws of motion,
the uniqueness of the earth. The list could be extended. Note that the
plenum does not logically necessitate either the uniqueness, or the
central position, or the immobility of the earth. It simply fits into a
coherent pattern in which the unique, central, and immobile earth is
a second essential strand. Conversely, the earth's motion does not
necessitate either the existence of a vacuum or the infinity of the universe. But it is no accident that both these views won acceptance
shortly after the victory of the Copernican theory.

Copernicus himself believed in neither. As we shall see, he tried to
preserve most of the central features of Aristotelian and Ptolemaic
cosmology. But by giving the earth an axial motion, he made the sphere
of the stars immobile, depriving it of physical function. And by giving
the earth an orbital motion, he made necessary a vast increase in the
size of the sphere. Copernicus' cosmology thus took away from inter-
planetary matter many of its essential Aristotelian functions and simultaneously demanded that there be vastly more of it. His successors soon
fractured the now functionless sphere, scattered the stars through all
of space, admitted a vacuum or something very like it between them,
and dreamed of other worlds inhabited by other men in the vast ex-
panses beyond our solar system. Even the terrestrial principle of the
horror vacui did not survive for long. In the new universe it was very
much easier for scientists to recognize that practical miners had for
a century been producing a terrestrial vacuum at the top of overlong
water pumps. Air pressure soon replaced the vacuum in the pneumatic conceptions of the seventeenth century. Many other forces
played an essential role in the modification of pneumatics — the story
is complex — but the new astronomy of Copernicus is a necessary in-

gredient of its plot. Once again astronomical theory displays its intimate entanglement with the theories of other sciences, and those other sciences condition the astronomical imagination.

The Majesty of the Heavens

The extra-astronomical entanglements of astronomical theory are not, however, exclusively ties to other sciences. As our previous discussions of the motives for celestial observation have repeatedly hinted, the ancient astronomical tradition is partially indebted for its very existence to a widespread primitive perception of the contrast between the power and stability of the heavens and the impotent insecurity of terrestrial life. This same perception is incorporated into Aristotle's cosmology by the absolute distinction between the superlunary and sublunary regions. But in Aristotle's highly articulated version the distinction comes to depend explicitly upon both the central position of the earth and the perfect symmetry of the spheres that generate both the stellar and planetary motions.

According to Aristotle, the underside of the sphere of the moon divides the universe into two totally disparate regions, filled with different sorts of matter and subject to different laws. The terrestrial region in which man lives is the region of variety and change, birth and death, generation and corruption. The celestial region is, in contrast, eternal and changeless. Only aether, of all the elements, is pure and incorruptible. Only the interlocked celestial spheres move naturally and eternally in circles, never varying their rate, always occupying exactly the same region of space, forever turning back upon themselves. The substance and the motion of the celestial spheres are the only ones compatible with the immutability and majesty of the heavens, and it is the heavens that produce and control all variety and change on earth. In Aristotle's physical description of the universe, as much as in any primitive religion, the encircling heavens are the locus of the perfection and the power upon which terrestrial life depends. *On the Heavens* puts the point unequivocally:

From what has been said it is clear why . . . the primary body of all [that is, celestial matter] is eternal, suffers neither growth nor diminution, but is ageless, unalterable and impassive. I think too that the argument bears out experience and is borne out by it. All men have a conception of gods, and all assign the highest place to the divine, both barbarians and Hellenes, as many as believe in gods, supposing, obviously, that immortal is closely linked with immortal. It could not, they think, be otherwise. If then — and it is true — there is something divine, what we have said about the primary bodily substance [namely that it is weightless, indestructible, unalterable, and so on] is well said. The truth of it is also clear from the evidence of the senses, enough at least to warrant the assent of human faith; for throughout all past time, according to the records handed down from generation to generation, we find no trace of change either in the whole of the outermost heaven or in any one of its proper parts. It seems too that the name of this first body has been passed down to the present time by the ancients. . . . Thus they, believing that the primary body was something different from earth and fire and air and water, gave the name *aether* to the uppermost region, choosing its title from the fact that it "runs always" and eternally.[6]

Aristotle himself carried the conception of the majesty and divinity of the celestial regions little further. Both the matter and the motions of the heavens are perfect; all terrestrial change is caused and governed by a succession of pushes initiated by the uniform motions of the celestial spheres which symmetrically enclose the earth. Already a significant nonscientific argument for the earth's unique central location is apparent, and, in the centuries after Aristotle's death, this argument was reinforced by elaborating the conception of the perfect heavens and integrating it with two other important sets of beliefs, both of which had originated independently. One of these developments — the detailed integration of Aristotelian cosmology with Christian theology — we must postpone until its proper chronological position in the next chapter. It resulted in a universe each of whose structural details carried religious as well as physical significance: Hell was at the geometric center; God's throne was beyond the stellar sphere; each planetary sphere and epicycle was turned by an angel. But another significant application of the concept of celestial majesty — the science of astrology — is older than the Christian-Aristotelian cosmology, and it had an even more immediate impact on practitioners of astronomy. Because it involved them professionally, astrology may well have been the most important of the forces binding astronomers to the conception of the earth's uniqueness.

We have already noted the principal roots of astrological belief and their relation to the Aristotelian conception of the power of the heavens. Distance and immutability make the heavens a plausible locus for the gods who can intervene at will in men's affairs. Disruptions of celestial regularity — particularly comets and eclipses — were regarded from an early date as portents of unusual felicity or disaster. In addition, there is good observational evidence for celestial control of at least some terrestrial events. It is hot when the sun is in the constellation Cancer and cold when it is in Capricorn; the varying height of the tide follows the variation of the moon's phases; the menstrual cycle of women throughout the earth recurs at intervals coincident with the length of the lunar month. In an era when man's need to understand and control his fate immeasurably transcended his physical and intellectual tools, this apparent evidence of celestial power was naturally extended to the other celestial wanderers. Particularly after Aristotle supplied a physical mechanism — the frictional drive — through which heavenly bodies could produce terrestrial change, there was a plausible basis for the belief that an ability to predict the future configurations of the heavens would enable men to foretell the future of men and nations.

Before the second century B.C., ancient records show few signs of a fully developed attempt to predict the details of terrestrial affairs from the observed and computed positions of the stars and planets. But after this relatively late start, astrology was inseparably linked to astronomy for 1800 years; together they constituted a single professional pursuit. The astrology that predicted the future of men from the stars was known as judicial astrology; the astronomy that predicted the future of stars from their present and past was known as natural astrology; those who gained fame in one branch were usually well known in the other as well. Ptolemy, whose *Almagest* exhibited ancient astronomy in its most developed form, was equally famous for his *Tetrabiblos*, antiquity's classic contribution to judicial astrology. European astronomers like Brahe and Kepler, who late in the Renaissance put Copernicus' system into something very like its modern form, were supported financially and intellectually because they were thought to cast the best horoscopes.

During most of the period with which the rest of this book is concerned, astrology exercised an immense influence upon the minds of the most educated and cultured men of Europe. Early in the Middle Ages it was partially suppressed by the Church, whose doctrinal insistence that men are free to choose the Christian good was incompatible with astrology's strict determinism. But during the five centuries centering on the birth of Christ, and again during the late Middle Ages and the Renaissance, astrology was the guide of kings and of their people, and it is no accident that these are just the periods during which earth-centered astronomy made most rapid progress. The elaborate tables of planetary position and the complex computational techniques developed by planetary astronomers from antiquity to the Renaissance were the main prerequisite for astrological prediction. Until after Copernicus' death these major products of astronomical research had little other socially significant application. Astrology therefore provided the principal motive for wrestling with the problem of the planets, so that astrology became a particularly important determinant of the astronomical imagination.

Astrology, however, and the perception of celestial power that underlies it lose much plausibility if the earth is a planet. A planetary earth will act as forcefully on Saturn as Saturn can act on it; the same argument applies to the other planets; and the terrestrial-celestial dichotomy breaks down. If the earth is a celestial body, it must share the immutability of the heavens, and the heavens in turn must participate in the corruptibility of earth. It cannot be coincidence that astrology's stranglehold upon the human mind finally relaxed during just the period in which the Copernican theory first gained acceptance. It may even be significant that Copernicus, the author of the theory that ultimately deprived the heavens of special power, belonged to the minority group of Renaissance astronomers who did not cast horoscopes.

Astrology and the majesty of the heavens therefore provide one more example of the indirect consequences of the earth's stability and uniqueness, consequences that have been repeatedly illustrated but by no means exhausted in this extended discussion of the multiple functions of a central stable earth in the Aristotelian world view. It is, of course, precisely these consequences and others like them that make the Copernican Revolution a revolution. To describe the innovation initiated by Copernicus as the simple interchange of the position of the earth and sun is to make a molehill out of a promontory in the development of human thought. If Copernicus' proposal had had no consequences outside astronomy, it would have been neither so long delayed nor so strenuously resisted.

The Aristotelian World View in Perspective

The Aristotelian world view was the single most important source and support for the pre-Copernican tradition of astronomical practice. But Aristotle's day is not our day, and a real mental transposition is therefore necessary in approaching his writings, particularly those dealing with physics and cosmology. Failure to make this transposition has resulted in some strained and distorted explanations of the endurance of Aristotelian physics in antiquity and during the Middle Ages.

We are, for example, often told that it is only because medieval scientists preferred the authority of the written word, preferably ancient, to the authority of their own eyes that they could continue to accept Aristotle's absurd dictum that heavy bodies fall faster than light ones. Modern science, on this prevalent interpretation, began when Galileo rejected texts in favor of experiments and observed that two bodies of unequal weight, released from the top of the tower of Pisa, struck the ground simultaneously. Today every schoolboy knows that heavy bodies and light bodies fall together. But the schoolboy is wrong and so is this story. In the everyday world, as Aristotle saw, heavy bodies do fall faster than light ones. That is the primitive perception. Galileo's law is more useful to science than Aristotle's, not because it represents experience more perfectly, but because it goes behind the superficial regularity disclosed by the senses to a more essential, but hidden, aspect of motion. To verify Galileo's law by observation demands special equipment; the unaided senses will not yield or confirm it. Galileo himself got the law not from observation, at least not from new observation, but by a chain of logical arguments like those we shall examine in the next chapter. Probably he did not perform the experiment at the tower of Pisa. That was performed by one of his critics, and the result supported Aristotle. The heavy body did hit the ground first.

The popular story of Galileo's refutation of Aristotle is largely a myth, motivated by a failure of historical perspective. We like to forget that many of the concepts in which we believe were painfully drummed into us in our youth. We too easily take them as natural and indubitable products of our own unaided perceptions, dismissing concepts different from our own as errors, rooted in ignorance or stupidity and perpetuated by blind obedience to authority. Our own education stands between us and the past. Particularly it stands between us and Aristotelian physics, frequently leading us to misinterpret the nature and source of Aristotle's immense influence on subsequent generations.

Part of the authority of Aristotle's writings derives from the brilliance of his own original ideas, and part derives from their immense range and logical coherence, which are as impressive today as ever. But the primary source of Aristotle's authority lies, I believe, in a third aspect of his thought, one which it is more difficult for the modern mind to recapture. Aristotle was able to express in an abstract and consistent manner many spontaneous perceptions of the universe that had existed for centuries before he gave them a logical verbal rationale. In many cases these are just the perceptions that, since the seventeenth century, elementary scientific education has increasingly banished from the adult Western mind. Today the view of nature held by most sophisticated adults shows few important parallels to Aristotle's, but the opinions of children, of the members of primitive tribes, and of many non-Western peoples do parallel his with surprising frequency. Sometimes the parallels are difficult to discover, because they are hidden by Aristotle's abstract vocabulary and by his elaborately logical method. These are the elements of Aristotelian dialectic, and only their rudiments can be found in primitives and children. But Aristotle's substantive ideas about nature, in contrast to the way he expresses and documents them, do show important residues of earlier and more elementary perceptions of the universe. Unless alert to these residues we may miss the meaning and will surely miss the force of important segments of Aristotelian doctrine.

The nature of the primitive residues and the manner in which they are transformed by the impact of Aristotelian dialectic are clearly illustrated in Aristotle's discussions of space and motion. The world views of primitive societies and of children tend to be animistic. That is, children and many primitive peoples do not draw the same hard and fast distinction that we do between organic and inorganic nature, between living and lifeless things. The organic realm has a conceptual priority, and the behavior of clouds, fire, and stones tends to be explained in terms of the internal drives and desires that move men and, presumably, animals. Asked why balloons go up, one child of four answers, "Because they want to fly away." Another, age six, explains that balloons go up because "they like the air. So when you let go they go up in the sky." Asked why a box falls to the ground, Hans, age five, answers, "Because it wants to go there." Why? "Because it is a good thing [for it to be there]." [7] Primitives frequently give similar explanations, though they are often harder to unravel because expressed in myths which cannot be taken quite literally. We have already examined the Egyptian explanation of the sun's motion as that of a god sailing in his boat across the heavens.

Aristotle's stones are not alive, though his universe frequently seems to be, at least metaphorically. (There are passages in Aristotle reminiscent of the passage from Plato's *Timaeus* quoted in the first chapter.) But his perception of the stone leaping from the hand to achieve its natural place at the center of the universe is not so very different from the child's perception of the balloon that likes the air or of the box that falls to the ground because it is good for it to be there. The vocabulary has changed; the concepts are manipulated with adult logic; animism has been transmuted. But much of the appeal of the Aristotelian doctrine must lie in the naturalness of the perception that underlies the doctrine.

Animism is not, however, the entire psychological base of Aristotle's explanation of motion. A subtler and, I think, more important element derives from the Aristotelian transmutation of a primitive perception of space. To the members of prehistoric civilizations and primitive tribes, space seems very different from the Newtonian space in which we were all brought up, usually without knowing it. The latter is physically neutral. A body must be located *in* space and move *through* space, but the particular part of space and the particular direction of motion exert no influence on the body. Space is an inert substratum for all bodies. Each position and each direction is like every other. In modern terminology, space is homogeneous and isotropic; it has no "top" or "bottom," "east" or "west."

The space of the primitive, in contrast, is often more nearly a life space: the space in a room, or in a house, or in a community. It has a "top" and "bottom," and "east" and "west" (or "front" and "back" — in many primitive societies words for direction derive from words for parts of the body and reflect the intrinsic differences of these parts). Each position is a position "for" some object or "where" some char-

acteristic activity occurs. Each region and direction of space is charac-
teristically different from every other, and the differences partially
determine the behavior of bodies in each region. Usually the primitive's
space is the active dynamic space of everyday life; distinct regions
have distinct characteristics.

Egyptian cosmology provided an example: the region of the circumpolar stars became the region of eternal life, of those who never
die. A similar perception of space provides one important source of
astrological thought. The nature and power of planets depends upon
their position in space. One old Babylonian text states: "When the
star Marduk [the planet Jupiter] is in the ascendant [that is, low on
the eastern horizon], it is then Nebo [the god Mercury]. When it has
risen . . . [number omitted] double hours, it is Marduk [the god
Jupiter]. When it stands in the middle of the heavens, it is Nibiru [the
highest one, the omnipotent god]. Each planet becomes this at its
zenith." [8]

The primitive residues inherent in the Aristotelian conception of
space are seldom so clear. But examine the following discussion of
motion from Aristotle's *Physics*:

> The typical locomotions of the elementary natural bodies — namely fire,
> earth, and the like — show not only that place is something, but also that it
> exerts a certain influence. Each is carried to its own place, if it is not
> hindered, the one up, the other down. . . . It is not every chance direction
> which is "up," but where fire and what is light are carried; similarly, too,
> "down" is not any chance direction but where what has weight and what
> is made of earth are carried — the implication being that these places do not
> differ merely in relative position, but also as possessing distinct potencies.[9]

This passage is an almost perfect summary of the conception of
space that underlies the Aristotelian explanation of motion: "place
. . . exerts a certain influence"; "places do not differ merely in relative position, but also as possessing distinct potencies." These are
places in a space that has an active and dynamic role in the motion of
bodies. Space itself supplies the push that drives fire and stones to
their natural resting places at the periphery and the center. The interactions of matter and space determine the motion and rest of bodies.
To us this is an unfamiliar notion, because we are the heirs of the
Copernican Revolution, which made it necessary to discard and re-

place the Aristotelian conception of space. But the concept is not implausible. Perhaps by mere coincidence, the spatial concepts embodied in Einstein's general theory of relativity are, in important respects, closer to Aristotle's than to Newton's. And Einstein's universe may, like Aristotle's and unlike Newton's, be finite.

Aristotle's world view was not the only one created in antiquity, nor was it the only one that gained adherents. But Aristotle's was far nearer to many primitive conceptions of the world than its ancient competitors, and it corresponded more closely with the evidence of unaided sense perception. That is another reason why it was so immensely influential, particularly during the late Middle Ages. Having isolated at least part of its appeal, we can better appreciate the strength that Aristotelian cosmology contributed to the ancient astronomical tradition. Now we must discover what happened to that tradition to prepare the way for Copernicus.

4

RECASTING THE TRADITION:

ARISTOTLE TO THE COPERNICANS

European Science and Learning to the Thirteenth Century

Aristotle was the last great cosmologist of antiquity, and Ptolemy, who lived almost five centuries after Aristotle, was its last great astronomer. Until after the death of Copernicus in 1543, the writings of these two men dominated the astronomical and cosmological thought of the West. Copernicus seems their immediate heir, for in the thirteen centuries that separate Ptolemy's death from Copernicus' birth no large and enduring modification had been imposed upon their work. Because Copernicus began where Ptolemy had stopped many people conclude that there was no science during the intervening centuries. In fact, there was much intense though spasmodic scientific activity, and it played an essential role in preparing the ground for the inception and success of the Copernican Revolution.

If there is a paradox here, it is only apparent. Thirteen centuries of spasmodic research did not materially modify the substantive beliefs of the researchers. Copernicus' teachers still believed that the structure of the universe was about as described by Aristotle and Ptolemy, and their beliefs place them in an ancient tradition. But their attitude toward those beliefs was not ancient. Conceptual schemes age with the succession of the generations that behold them. At the beginning of the sixteenth century men still believed in the ancient description of the universe, but they evaluated it differently. Their concepts were the same, but they saw new strengths and weaknesses in those concepts. Just as we have explored the sources and the strengths of the ancient astronomical tradition, so we must discover what happened to it as it aged. We shall have to begin by finding out how the ancient tradition was lost and then rediscovered, for the first changes in European attitudes toward the tradition arose from the necessity of recovering it.

The West's loss of ancient science occurred in two stages, the first a slow decline in the quality and quantity of scientific activity, the second a genuine disappearance of traditional learning. After the second century B.C., Mediterranean civilization was increasingly dominated by Rome, and it declined with the decline of Roman hegemony during the first few centuries of the Christian era. Ptolemy, the astronomer, and Galen, the physician, were the last great figures in ancient science, and they both lived in the second century A.D. After their time the major scientific works of the West were commentaries and encyclopedias. When the Moslems invaded the Mediterranean basin during the seventh century, they found only the documents and the tradition of ancient learning. The activity had largely ceased.

The Islamic invasions shifted the geographic center of European Christendom northward from the Mediterranean and thus enforced the continued decline of Western learning. During the seventh century Europeans were deprived even of the documents in which the ancient learned tradition was embodied. Euclid was known only in the incomplete Latin translation prepared by Boethius early in the sixth century; that version stated only the more important theorems and included no proofs. Ptolemy was apparently totally unknown, and Aristotle was represented only by a few works on logic. Encyclopedic collections by men like Boethius and Isidore of Seville preserved fragments of ancient science, but even these fragments were too often inaccurate, intellectually debased, and heavily interlarded with fable. There was little learned activity of any sort. The economic level of European Christendom barely provided subsistence. Science was particularly neglected because, as we shall see in the next section, the Catholic Church was initially hostile to it.

During the centuries when European learning reached its nadir, there was a great renaissance of science in Islam. After the middle of the seventh century the Moslem world rapidly expanded from an Arabian oasis to a Mediterranean empire, and this new empire inherited the manuscripts and the tradition that Christendom had lost. Moslem scholars first reconstituted ancient science by translating Syriac versions of original Greek texts into Arabic; then they added contributions all their own. In mathematics, chemistry, and optics they made original and fundamental advances. To astronomy they contributed both new observations and new techniques for the computation of planetary position. Yet the Moslems were seldom radical innovators in scientific theory. Their astronomy, in particular, developed almost exclusively within the technical and the cosmological tradition established in classical antiquity. Therefore, from our present restricted viewpoint, Islamic civilization is important primarily because it preserved and proliferated the records of ancient Greek science for later European scholars. Christendom recovered ancient learning first from the Arabs and usually in Arabic translation. The title *Almagest* by which we know Ptolemy's major work is not its Greek name at all, but a contraction of the Arabic title which it received from a ninth-century Moslem translator.

Europeans rediscovered ancient learning in Islam during the period of general European recovery which makes the tenor of the later Middle Ages so different from that of the Dark Ages. Beginning slowly in the tenth century and culminating in what is now known as the Twelfth Century Renaissance, there was a gradual increase in the tempo of all aspects of European life. For the first time Christendom achieved relative political security; with it came a great increase of population and trade, including trade with the Moslem world. Intellectual contacts with Islam multiplied as commerce grew. New-found wealth and security provided leisure to explore the newly opened horizons of learning. The first Latin translations from the Arabic were made in the tenth century and multiplied rapidly in those that followed. Late in the eleventh century students from all over Europe began to assemble informally but in steadily increasing numbers to hear some master read and comment upon a new translation of an ancient text. In the twelfth and thirteenth centuries these initially informal gatherings became so large that they required the regulations and charters that transformed them officially to universities, a new sort of learned institution indigenous to Europe. Starting as centers for the oral propagation of ancient learning, these universities rapidly became the home of an original and creative tradition of European scholarship, the critical and combative philosophical tradition known as scholasticism.

The rediscovery of ancient astronomy was a part of the larger reclamation of the science and philosophy of the ancient world. The first astronomical tables to be widely exploited by Europeans were imported from Toledo in the eleventh century. Ptolemy's *Almagest* and most of Aristotle's astronomical and physical writings were latinized during the twelfth, and in the following century they were steadily, though selectively, integrated into the curriculum of the medieval university. Copernicus studied them there at the end of the fifteenth century, and his return to these classics of ancient science makes him the heir of Aristotle and Ptolemy. But they would scarcely have recognized as their own work the inheritance that Copernicus received. Old problems, though still unsolved, had disappeared; new ones, though sometimes merely pseudo problems, had taken their place. In addition, the purposes and methods of the revived learned tradition differed significantly from the ones that had guided ancient scholars.

Some of the new problems were purely textual in origin. Ancient texts were recovered piecemeal, in an order governed more by chance than by logic. Arabic manuscripts were seldom completely faithful to their Greek or Syriac sources; the medieval Latin into which they were retranslated did not at first possess a vocabulary adequate to their technical and abstract subject matter; even good translations inevitably deteriorated during successive transcriptions by men who did not fully understand them. Discovering how Aristotle or Ptolemy had answered a particular question was often difficult and sometimes impossible. Yet medieval scholars repeatedly insisted on reconstructing ancient thought before venturing a judgment of their own. The brilliance, scope, and coherence of their unexpected legacy dazzled men emerging from the Dark Ages; they naturally felt that their first task was to assimilate their heritage. Problems of interpretation and reunification therefore bulked large in scholastic thought.

The task of the medieval scholar was further and artificially complicated by a foreshortened historical perspective. He expected to reëstablish a broad and coherent system of knowledge modeled on Aristotle's, and he did not always recognize that the "antiquity" from whom the system was to derive had had a number of different opinions about a great many questions of detail. Though the scholastics found it hard to recognize (often pleading errors of transmission or translation), Aristotle himself had not always been consistent. Nor had his contemporaries accepted all his views. Occasional equivocations and contradictions had characterized the ancient tradition from the start. Their range had been immensely broadened by the Hellenistic and Islamic commentators, whose works, written during the fifteen centuries separating Aristotle from his European disciples, were recovered with, and sometimes before, those of the master. To us these inconsistencies in the tradition seem natural products of its evolution and transmission, but to the medieval scholar they often appeared as internal contradictions in a single body of knowledge, the hypothetical unit "ancient wisdom." In part because of this confusion, the comparison and reconciliation of conflicting authorities became a distinctive characteristic of scholastic thought. As we shall see more concretely later in this chapter, the revived tradition of learning was less empirical, more verbal, logical, and rational, than its ancient counterpart had been.

One of the inconsistencies embedded in the tradition played a particularly significant role in the development of astronomy: the apparent conflict between the spheres of Aristotelian cosmology and the epicycles and deferents of Ptolemaic astronomy. Though we have not previously noted it, these were really characteristic products of two distinct ancient civilizations, the Hellenic and the Hellenistic. Hellenic civilization centered on the Greek mainland during the period when Greece dominated the Mediterranean basin. The science to which it gave rise was predominantly qualitative in method and cosmological in orientation. Aristotle was its greatest representative and also its last. Just before his death the evolution of Hellenic science came to a premature halt at the time when Alexander the Great conquered Greece and joined it to a great empire embracing all of Asia Minor, Egypt, and Persia to the Indus river. The Hellenistic civilization that emerged after Alexander's conquests centered in commercial and cosmopolitan metropolises like Alexandria. There scholars of many nations and races merged elements of their diverse cultures to produce a science that was less philosophical, more mathematical and numerical, than its Hellenic predecessor had been. Astronomy perfectly illustrates the contrast. The cosmological framework of ancient astronomy is largely a product of the Hellenic tradition which culminated in the works of Aristotle. The mathematical astronomy of Hipparchus and Ptolemy belongs to the Hellenistic tradition which, in astronomy, flourished only two centuries and more after Aristotle's death.

The Hellenistic astronomers who measured the universe, catalogued the stars, and grappled with the problem of the planets were clearly not indifferent to the cosmology developed by their Hellenic predecessors. But neither were they much concerned with cosmological minutiae. They ridiculed the authors of cosmologies that differed radically from the established norm, and they occasionally wrote short cosmological treatises of their own. Ptolemy himself composed a thoroughly cosmological work, the *Hypotheses of the Planets*, which includes a rather unsatisfactory physical mechanism for epicyclic motions. But when designing mathematical systems to predict planetary position Hellenistic astronomers seldom worried about the possibility of constructing mechanical counterparts for their geometric constructs. To them the physical reality of the spherical shells and the mechanisms which kept the planets moving within them had become at most secondary problems. In short, Hellenistic scientists acquiesced without apparent discomfort in a partial bifurcation of astronomy and cosmology; a satisfactory mathematical technique for predicting planetary position did not have to conform entirely to the psychological requirement of cosmological reasonableness.

In the sixteenth century this bifurcation offered an important precedent to Copernicus. Because he too saw astronomy as essentially mathematical, the physical incongruity of a moving epicycle in a universe of spheres could provide a dim anticipation of the physical incongruity of a moving earth. But this was not the bifurcation's first contribution. Four centuries earlier, at the time when Aristotle and Ptolemy were first recovered by Europeans, it had also helped pave the road to revolution, though in a very different way. Because their ignorance of the preceding centuries had telescoped their sense of history, the scholastics viewed Aristotle and Ptolemy very nearly as contemporaries. They appeared as exponents of a single tradition, "ancient learning," and the differences between their systems became very like inconsistencies in a single body of doctrine. Changes that Ptolemy had seen as the natural evolution of knowledge over the five centuries separating him from Aristotle, often seemed contradictions to the scholastics, and contradictions raised novel problems of reconciliation. Since the passage of time proved reconciliation to be both difficult and inconclusive, these apparent contradictions, like other conflicts in medieval thought, finally helped to cast doubt upon the entire tradition.

As revived in the Middle Ages, the ancient tradition of learning had acquired a new look, and the preceding pages indicate that some of the important novelties derived from the mere necessity of revival. But there were also more substantive changes in the revived tradition, changes produced by indigenous characteristics of the Middle Ages and the Renaissance. For example, though science played a large part in the thought of the later Middle Ages, the dominant intellectual forces were theological, and the practice of science in a theological milieu shifted both the strengths and the weaknesses of the scientific tradition. Besides, medieval science was not itself static. Aristotle's scholastic critics developed important alternatives for some of his doctrines, and a few of these alternatives played a major role in preparing the way for Copernicus. By the sixteenth century still other forces — intellectual, economic, and social — were at work, and among these are some with a direct bearing upon the problems of astronomy and of the earth's motion. These changes demand an independent treatment, to which we now turn.

Astronomy and the Church

Throughout the Middle Ages and much of the Renaissance the Catholic Church was the dominant intellectual authority of all Europe. Medieval European scholars were members of the clergy; the universities in which ancient learning was assembled and studied were Church schools. From the fourth century to the seventeenth the Church's attitude toward science and about the structure of the universe was a determining factor in the progress or stagnation of astronomy. But the Church's attitude and its practice were not uniform during these centuries. After the Dark Ages the Church began to support a learned tradition as abstract, subtle, and rigorous as any the world has known. But before the tenth century and again after the sixteenth the Church's influence was, on balance, antiscientific. The Copernican theory evolved within a learned tradition sponsored and supported by the Church; Copernicus himself was the nephew of a bishop and a canon of the cathedral at Frauenburg; yet in 1616 the Church banned all books advocating the reality of the earth's motion.

No single generalization will describe the Church's overwhelming influence upon science, for the influence changed with the changing situation of the Church.

In the early centuries of the Christian era the Church fathers were crusaders and proselytizers for a new faith, fighting for its very existence. Their calling itself demanded that they deprecate the pagan learning of their predecessors and maximize the attention given to the problems of Christian theology by the rapidly contracting learned world. In addition, they deeply believed that Scripture and Catholic exegesis contained the sum of knowledge necessary for salvation. To them science was secular learning. Except when essential for daily life it was useless at best, dangerously distracting at worst. Therefore, in his *Enchiridion* or handbook for Christians, St. Augustine, the most influential of the early Church fathers, counseled the faithful as follows:

> When, then, the question is asked what we are to believe in regard to religion, it is not necessary to probe into the nature of things, as was done by those whom the Greeks call *physici*; nor need we be in alarm lest the Christian should be ignorant of the force and number of the elements, — the motion, and order, and eclipses of the heavenly bodies; the form of the heavens; the species and the natures of animals, plants, stones, fountains, rivers, mountains; about chronology and distances; the signs of coming storms; and a thousand other things which those philosophers either have found out, or think they have found out. . . . It is enough for the Christian to believe that the only cause of all created things, whether heavenly or earthly, whether visible or invisible, is the goodness of the Creator, the one true God; and that nothing exists but Himself that does not derive its existence from Him.[1]

This attitude was not incompatible with an admiring knowledge of ancient learning, at least not before the Moslem invasions. Augustine himself had read Greek science attentively, and his writings testify to his admiration of its accuracy and scope. But his attitude was incompatible with the active pursuit of scientific problems, and it readily lent itself to further negative elaboration. In the writings of Augustine's less liberal contemporaries and successors, his spiritual depreciation of pagan science was usually coupled with an outright rejection of its content. Astronomy, because of its ties to astrology, was particularly scorned, for the explicit determinism of astrology made it seem incompatible with Christian doctrine.

At the beginning of the fourth century, for example, Lactantius, tutor to the son of the Emperor Constantine, devoted the third book of his *Divine Institutions* to "the false wisdom of the philosophers" and allowed himself one chapter to ridicule the concept of the spherical earth. For him it was sufficient to point to the absurdity of a region in which men hung head down and to the impossibility of the heavens' being below the earth. Later in the century the Bishop of Gabala achieved the same effect with Biblical evidence. The heavens are not a sphere, but a tent or tabernacle, for "it is He . . . that stretcheth out the heavens as a curtain, and spreadeth them out as a tent to dwell in" (Isaiah 40:22). There are "waters . . . above the firmament" (Genesis 1:7). The earth is flat, for "the sun was risen upon the earth when Lot entered into Zoar" (Genesis 19:23). By the middle of the sixth century Kosmas, an Alexandrian monk, could replace the pagan system with a detailed Christian cosmology derived primarily from the Bible. His universe is shaped like the tabernacle that the Lord instructed Moses to build in the wilderness. It has a flat bottom, perpendicular sides, and a semicylindrical roof, like an old-fashioned traveling trunk. The earth, the footstool of the Lord, is a rectangular plane, twice as long as it is broad, and resting on the flat bottom of the universe. The sun does not travel below the earth at night, but is hidden behind its northernmost portions, which are higher than the regions to the south.

The cosmologies of men like Lactantius and Kosmas never became official Church doctrine. Nor did they entirely supplant the ancient universe of spheres which survived in fragmentary descriptions within the more erudite medieval encyclopedias. There was no Christian unanimity about cosmology during the first half of the Middle Ages. Science and cosmology were not significant enough to demand it. But though these cosmologies, compounded from the naïvest sense perceptions plus a smattering of Scripture, were never official, they are representative. They illustrate the decadence of secular learning that characterized the Dark Ages, and they therefore prepare us for the surprise and awe with which later Christian scholars greeted the rediscovery of ancient knowledge during the eleventh and twelfth centuries.

By the time that Christian Europe reëstablished commercial and cultural ties with the Eastern Church in Byzantium and with the Moslems of Spain, Syria, and Africa, the Church's attitude toward pagan wisdom had changed. The main areas of continental Europe had been converted; the Church's intellectual and spiritual authority was complete; the hierarchy of ecclesiastical administration was fixed. Pagan and secular learning were no longer a threat, provided that the Church could maintain intellectual leadership by absorbing them. Churchmen therefore devoted some of the leisure provided by new-found prosperity to a vigorous pursuit of the rediscovered learning. By broadening the range of knowledge acceptable for Christian scholarship, they preserved for five more centuries the Catholic monopoly of learning. In the twelfth century "the nature of things," including the heavens and the earth, again became a suitable topic for intensive study. By the thirteenth, if not earlier, the main outlines of the two-sphere universe were once more taken for granted in the discussions of educated Christians. During the last centuries of the Middle Ages the setting of Christian life, both terrestrial and celestial, was a fully Aristotelian universe.

We have been calling the process by which Christians discovered that they lived in an Aristotelian universe a recovery of ancient learning, but "recovery" is clearly an inadequate word. What occurred was far more nearly a revolution in both Christian thought and the ancient scientific tradition. From the fourth century on, Aristotle, Ptolemy, and other Greek writers had been attacked by Churchmen because of the conflict between their cosmological opinions and Scripture. Those conflicts still existed in the twelfth and thirteenth centuries, and they were recognized. In 1210 a provincial council at Paris prohibited the teaching of Aristotelian physics and metaphysics. In 1215 the Fourth Lateran Council issued a similar, though more restricted, anti-Aristotelian edict. Other interdictions issued from the papacy throughout the century. They were unsuccessful, winning lip service alone, but they are not insignificant. The edicts testify to the impossibility of simply adding ancient secular learning to the existing body of medieval theology. Both ancient texts and Scripture required modification in the creation of a new fabric of coherent Christian doctrine. When the new fabric was completed, theology had become an important bulwark for the ancient concept of a central stationary earth.

The physical and cosmological structure of the new Christian universe was predominantly Aristotelian. St. Thomas Aquinas (1225–1274), the scholastic who contributed most to the final pattern of the fabric, describes the perfection and appropriateness of the celestial motions in words that, except for their clarity, might have been written by Aristotle himself:

It is therefore clear that the material of the heavens is, by its intrinsic nature, not susceptible to generation and corruption, since it is the primary sort of alterable body and closest in its nature to those bodies which are intrinsically changeless. [The only truly changeless body in the Christian universe is God, from whom all change on earth and in the heavens derives.] That is why the heavens experience only the absolute minimum of alteration. Motion is the only sort of change they experience, and this sort of alteration [unlike change of size, weight, color, and so on] does not modify their intrinsic nature in the least. Furthermore, among the sorts of motion which they might experience, theirs is circular, and circular motion is the one which produces the very minimum of alteration because the sphere as a whole does not change place.[2]

Aristotle could not always be embraced quite so literally. Many scholastics felt forced, for example, to abandon his proof of the absolute impossibility of a void, because it seemed arbitrarily to limit God's infinite power. No Christian could accept Aristotle's view that the universe had always existed. The first words of the Bible are, "In the beginning God created the heaven and the earth." Besides, the Creation was an essential ingredient in the Catholic explanation of the existence of evil. On a matter of this significance Aristotle had to give way; the universe had been created at a determinate first instant of time. But more often the Bible gave way, usually to metaphorical interpretation. For example, in discussing the scriptural text, "Let there be a firmament made amidst the waters; and let it divide the waters from the waters" (Genesis 1:6), Aquinas first outlined a cosmological theory that would preserve the literal sense of the passage and then said:

As, however, this theory can be shown to be false by solid reasons, it cannot be held to be the sense of Holy Scripture. It should rather be considered that Moses was speaking to ignorant people, and that out of condescension to their weakness he put before them only such things as are apparent to sense. Now even the most uneducated can perceive by their senses that earth and water are corporeal, whereas it is not evident to all that air also is corporeal. . . . Moses, then, while he expressly mentions water and earth, makes no express mention of air by name, to avoid setting before ignorant persons something beyond their knowledge.[3]

By reading "water" as "air" or "transparent substance" the integrity of Scripture is preserved. But in the process the Bible becomes, in some sense, a propaganda instrument, composed for an ignorant audience. The device is typical, the scholastics employed it again and again.

The painstaking thoroughness with which Aquinas and his contemporaries attacked the task of reconciliation is illustrated by the difficulties they discovered in the Biblical account of the Ascension. According to Scripture Christ "ascended up far beyond all heavens, that he might fill all things" (Ephesians 4:10). Aquinas succeeded in fitting this bit of Christian history into a universe of spheres, but to do so he had to resolve many varied problems, among which were the following:

It seems that it was not fitting for Christ to ascend into heaven. For the Philosopher [Aristotle] says (*On the Heavens*, Book II) that *things which are in a state of perfection possess their good without movement.* But Christ was in a state of perfection. . . . Consequently, He has His good without movement. But ascension is movement. Therefore it was not fitting for Christ to ascend. . . .

Further, there is no place above the heavens, as is proved in *On the Heavens*, I. But every body must occupy a place. Therefore Christ's body did not ascend above all the heavens. . . .

Further, two bodies cannot occupy the same place. Since, then, there is no passing from place to place except through the middle space, it seems that Christ could not have ascended above all the heavens unless [the crystal spheres of] heaven were divided; which is impossible.[4]

Aquinas' answers need not concern us. It is the objections themselves that are astounding, particularly since the Ascension is only one of the aspects of Christian history to present problems and since Aquinas is only the greatest of the many Churchmen concerned with them. Aquinas' *Summa Theologica*, from which most of the preceding quotations were taken, is a compendium of Christian knowledge, often printed in twelve fat volumes. In all of them Aristotle's name (or his more revealing designation as "The Philosopher") occurs again and again. Only through a multiplicity of works like these could ancient learning, particularly Aristotelian learning, have again become a foundation for Western thought.

Aquinas and his thirteenth-century contemporaries certified the compatibility of Christian belief with much of ancient learning. By making Aristotle orthodox they licensed his cosmology to become a creative element in Christian thought. But the very detail and erudition of their works obscured the over-all structure of the new Christian universe that was emerging late in the Middle Ages. If we are to understand the pregnant meaning which gave that universe, including the central, stable earth, its hold upon the medieval and Renaissance mind, we require a more comprehensive view. That larger view can scarcely be discovered in the thirteenth century at all. It evolved only after Aristotle had been licensed, appearing perhaps first and certainly most forcefully in the works of the Italian poet Dante, particularly in his great epic, the *Divine Comedy*.

Taken literally, Dante's epic is a description of the poet's journey through the universe as conceived by the fourteenth-century Christian. The journey begins on the surface of the spherical earth; descends gradually into the earth via the nine circles of Hell which symmetrically mirror the nine celestial spheres above; * and arrives at the vilest and most corrupt of all regions, the center of the universe, the appropriate locus of the Devil and his legions. Dante then returns to the surface of the earth at a point diametrically opposite the one where he had entered, and there he finds the mount of purgatory with its base on the earth and its top extending into the aerial regions above. Passing through purgatory, the poet travels through the terrestrial spheres of air and fire to the celestial region above. At the last he journeys through each of the celestial spheres in turn, conversing with the spirits that inhabit them, until finally he contemplates God's Throne in the last, the Empyrean, sphere. The setting of the *Divine Comedy* is a literal Aristotelian universe adapted to the epicycles of Hipparchus and the God of the Holy Church.

For the Christian, however, the new universe had symbolic as well as literal meaning, and it was this Christian symbolism that Dante wished most of all to display. Through allegory his *Divine Comedy* made it appear that the medieval universe could have had no other structure than the Aristotelian-Ptolemaic. As he portrays it, the universe of spheres mirrors both man's hope and his fate. Both physically

---

* The ninth sphere, which appears throughout medieval astronomy, was added to the eight spheres of ancient cosmology by Moslem astronomers in order to account for the precession of the equinoxes, and the motion of the pole of the heavens (see the Technical Appendix, Section 2). In the Moslem system it is the ninth sphere that rotates every 24 hours, as the sphere of the stars had done in the older system.

and spiritually man occupies a crucial intermediate position in this universe filled, as it is, by a hierarchical chain of substances that stretches from the inert clay of the center to the pure spirit of the Empyrean. Man is compounded of a material body and a spiritual soul: all other substances are either matter or spirit. Man's location, too, is intermediate: the earth's surface is close to its debased and corporeal center but within sight of the celestial periphery which surrounds it symmetrically. Man lives in squalor and uncertainty, and he is very close to Hell. But his central location is strategic, for he is everywhere under the eye of God. Both man's double nature and his intermediate position enforce the choice from which the drama of Christianity is compounded. He may follow his corporeal, earthy nature down to its natural place at the corrupt center, or he may follow his soul upward through the successively more spiritual spheres until he reaches God. As one critic of Dante has put it, in the *Divine Comedy* the "vastest of all themes, the theme of human sin and salvation, is adjusted to the great plan of the universe." [5] Once this adjustment had been achieved, any change in the plan of the universe would inevitably affect the drama of Christian life and Christian death. To move the earth was to break the continuous chain of created being.

No aspect of medieval thought is more difficult to recapture than the symbolism that mirrored the nature and fate of man, the microcosm, in the structure of the universe, which was the macrocosm. Perhaps we can no longer grasp the full religious significance with which this symbolism clothed the Aristotelian spheres. But we can at least avoid dismissing it as mere metaphor or supposing that it had no active role in the Christian's nonastronomical thought. One of Dante's prose works, written in part as a technical handbook to aid his contemporaries in interpreting his verse, concludes a literal physical description of the spheres and epicycles employed in medieval astronomy with the following words:

However, beyond all these [crystalline spheres], the Catholics place the Empyrean Heaven . . . ; and they hold it to be immovable, because it has within itself, in every part, that which its matter demands. And this is the reason that the *Primum Mobile* [or ninth sphere] moves with immense velocity; because the fervent longing of all its parts to be united with those of this most quiet heaven, makes it revolve with so much desire that its velocity is almost incomprehensible. And this quiet and peaceful heaven is the abode of that Supreme Deity who alone doth perfectly behold himself. [6]

In this passage the astronomer charts the position (and elsewhere the dimensions) of God's abode. He has become theologian, and in the fourteenth and fifteenth centuries the astronomer's theological functions did not always end with the measurement of heaven. Dante and some of his contemporaries also turned to astronomy to discover the kinds and occasionally even the numbers of the angelic inhabitants of God's spiritual realm.

Dante himself outlines one typical medieval theory of the relation between the spiritual hierarchy and the spheres in a passage from *The Banquet* immediately following the description of the spheres referred to above:

Since it has been demonstrated in the preceding chapter what this . . . heaven is, and how it is ordered within itself, it remains to show who they are who move it. Therefore be it known, in the first place, that these are Substances separate from matter, that is, Intelligences, whom the common people call Angels. . . . The numbers, the orders, the hierarchies [of these angelic beings], are recounted by the movable heavens, which are nine; and the tenth announces the unity and stability of God. And therefore the psalmist says, "The heavens recount the glory of God, and the firmament announceth the work of His hands."

Wherefore it is reasonable to believe that the motive powers [that is, the beings who move the spheres] of the Heaven of the Moon are of the order of Angels; and those of Mercury, Archangels; and those of Venus are the Thrones. . . . And these Thrones, which are allotted to the government of this heaven [of Venus], are not many in number, and the astrologers [or astronomers] differ about their number, according to their differences about the revolutions [of this heaven], although all are agreed in this, that their number is equal to that of these revolutions; which, according to the *Book of the Aggregation of the Stars* . . . are three: one by which the star revolves within its epicycle, the second by which the epicycle and the whole heaven revolves equally with that of the Sun, and the third by which all that heaven revolves, following the [precessional] motion of the stellar sphere from west to east, one degree in a hundred years. So that for these three motions are three motive powers [which are three members of the angelic order of Thrones].[7]

When angels become the motive force of epicycles and deferents, the variety of spiritual creatures in God's legion may increase with the complexity of astronomical theory. Astronomy is no longer quite separate from theology. Moving the earth may necessitate moving God's Throne.

Aristotle's Scholastic Critics

The effects of medieval scholarship were not all so conservative as the integration that made theology a bulwark for the universe
of spheres. Aristotle and his commentators were the invariable starting
point for scholastic research, but they were often no more than that.
The very intensity with which Aristotelian texts were studied
guaranteed that inconsistencies of doctrine or proof would be quickly
noticed, and these inconsistencies were often the seeds of important
creative achievements. Medieval scholars caught scarcely a glimpse of
the new astronomy and cosmology produced by their sixteenth- and
seventeenth-century successors. But they extended Aristotle's logic,
discovered fallacies in his proofs, and rejected many of his explanations
because they failed the test of experience. In the process they forged
a number of the concepts and tools that proved essential to the ac-
complishments of men like Copernicus and Galileo.

Important anticipations of Copernican thought can be found, for
example, in the critical commentary on Aristotle's *On the Heavens*
written during the fourteenth century by Nicole Oresme, a member of
the important Parisian nominalist school. Oresme's method is typically
scholastic. In his long manuscript Aristotle's text is broken into fragments, each a few sentences in length, and these fragments are inter-
spersed with long critical or explanatory comments. On finishing the
work a reader discovers that Oresme agrees with Aristotle on almost
every substantive point except the Creation. But the reasons for his
agreement are far from clear; Oresme's brilliant critique has destroyed
many of Aristotle's proofs and suggested important alternatives for a
number of Aristotelian positions. These alternatives were seldom
adopted by the scholastics themselves. But medieval scholars continued to discuss them, and that discussion helped to create a climate
of opinion in which astronomers could imagine experimenting with
the idea of a moving earth.

Oresme was, for example, quite critical of Aristotle's principal argument for the earth's uniqueness.[8] Aristotle had said that if there were
two earths in space (and when the earth becomes a planet, there will
be six "earths"), then they would both fall to the center of the universe
and coalesce, because earth moves naturally to the center. This proof,
says Oresme, is invalid, because it presupposes a theory of motion that

is itself unproved. Perhaps earth does not move naturally to the center, but simply to other nearby bits of earth. Our earth has a center, and it may be to that center, wherever in the universe it is located, that loose stones return. On this alternative theory the natural motion of a body is governed, not by its position in an absolute Aristotelian space, but by its position relative to other portions of matter. Some such theory was prerequisite to the new cosmologies of the sixteenth and seventeenth centuries, cosmologies in which the earth was neither unique nor at the center. Similar theories, in various disguises, are common to the work of Copernicus, Galileo, Descartes, and Newton.

Even more important anticipations of Copernican arguments emerge when Oresme criticizes Aristotle's refutation of Heraclides, the Pythagorean who had explained the diurnal motion of the stars by positing an eastward axial rotation of the central earth. Oresme does not believe that the earth rotates, at least he says he does not. But he is concerned to show that the choice between a stable and a rotating earth must be made on faith. No argument, he says, whether logical, physical, or even scriptural, can disprove the possibility of the earth's diurnal rotation. For example, nothing can be concluded from the apparent motion of the stars, because, says Oresme:

I suppose that local motion can be perceived only when one body alters its position relative to another. Therefore if a man in a smoothly riding boat, *a*, which is moved either slowly or rapidly, can see nothing but a second boat, *b*, which moves in just the same way as *a*, . . . then I say that it will seem to him that neither boat is moved. And if *a* is at rest and *b* moves, it will seem to him that *b* moves; and if it is *a* that moves and *b* is at rest, it will still seem to him that *a* is at rest and that *b* is moved. . . . Therefore I say that if the higher [or celestial] of the two parts of the universe mentioned above were today moved with a diurnal motion, as it is, while the lower [or terrestrial] part remained at rest, and if tomorrow on the contrary the lower part were moved diurnally while the other part, i.e., the heavens, were at rest, we would be unable to see any change, but everything would seem the same today and tomorrow. It would seem to us throughout that our location was at rest while the other part of the universe moved, just as it seems to a man in a moving boat that the trees outside the boat are in motion.[9]

This is the argument from optical relativity that plays such a large part in the writings of Copernicus and Galileo. Oresme does not stop with it, however. His treatise proceeds immediately to the demolition of an even more important Aristotelian argument, the one that concludes for immobility because an object thrown vertically upward always returns to the point on earth from which it departed:

> [In response to Aristotle's and Ptolemy's argument] one may say that an arrow shot straight into the air is [also] moved rapidly eastward with the air through which it passes and with the whole mass of the bottommost [or terrestrial] portions of the universe described above, the whole [earth and air and arrow] being moved with a daily rotation. Therefore the arrow returns to the spot on the earth from which it was shot. This appears possible by analogy: if a man were on a ship moving rapidly eastward without his being aware of its motion, and if he drew his hand rapidly downward, describing a straight line against the mast of the ship, it would seem to him that his hand had only a vertical motion; and the same argument shows why the arrow seems to us to go straight up or down.[10]

Galileo's famous defense of the Copernican system, the *Dialogue on the Two Principal Systems of the World*, is filled with arguments of exactly this sort, and Galileo may well have been elaborating hints derived from Copernicus' scholastic predecessors, including Oresme. But that does not make Oresme a Copernicus. He does not conclude for even the diurnal rotation of the earth; he does not dream of an orbital motion about the center of the universe; and he has no notion of the benefits that astronomers might derive by positing a mobile earth. He does not, for that matter, even share Copernicus' motive, and its absence makes his work all the more astounding. When Oresme's arguments recur in the writings of Copernicus and Galileo, they have a different and a more creative function. The later scientists wished to show that the earth *could move* in order to exploit the advantages that would accrue to astronomy if, in fact, it *did move.* Oresme wished to show only that the earth *could move*; he was investigating only Aristotle's proof. Like many of the other most fruitful contributions of scholastic science, his Copernican arguments are products of the preëminence that the late medieval mind accorded Aristotle. Men who agreed with Aristotle's conclusions investigated his proofs only because they were proofs executed by the master. Nevertheless their investigations often helped to ensure the master's ultimate overthrow.

We cannot, of course, be certain that Copernicus or Galileo read Oresme. The tradition that requires the scholar and scientist to name his sources was not established until long after the scientific revolution of the sixteenth and seventeenth centuries. But Aristotle had many scholastic critics, and they wrote numerous manuscripts which were repeatedly copied in the years after their deaths. Five and a half centuries after the composition of Oresme's commentary there are still six extant medieval manuscript copies, several dating from the fifteenth century, after Oresme's death. At the time of Copernicus' birth there must have been many more. Furthermore, the tradition of scholastic criticism was a continuous tradition. Key concepts which originated at Paris in the fourteenth century can be traced to Oxford in the same century and to Padua in the fifteenth and sixteenth. Copernicus studied at Padua and Galileo taught there. Though we cannot be sure that Copernicus derived any particular argument in the *De Revolutionibus* from any particular scholastic critic, we cannot doubt that the critics as a group facilitated the production of those arguments. At the very least they created a climate of opinion in which topics like the earth's motion were legitimate subjects for university discussion. Quite probably a few of Copernicus' key arguments were borrowed from earlier and unacknowledged sources.

The preceding discussion of Oresme has illustrated the most typical sort of scholastic criticism: the testing of Aristotelian proofs and the investigation of possible alternative doctrines, usually discarded once their logical possibility was demonstrated. But not all medieval science was of this limited critical and perhaps evanescent variety. The scholastics also introduced a few new areas of investigation and a few permanent doctrinal modifications into the Aristotelian scientific tradition. Among these the most significant were in the fields of kinematics and dynamics whose subject is the motion of heavy bodies on the earth and (after the Middle Ages) in the heavens. Some of Galileo's most significant contributions, particularly his work on falling bodies, can appropriately be viewed as a creative reordering of previously scattered physical and mathematical insights gained with difficulty by medieval scholars. But even before the seventeenth century, when Galileo wove them into a new dynamics, one of these insights, the impetus theory of motion, had had an important, if indirect, bearing on astronomical thought.

The impetus theory was erected on the rubble of one of the weakest explanations in the body of Aristotle's physics, the explanation of projectile motion. Aristotle had believed that, unless it was moved by an external push, a stone would either remain at rest or move in a straight line toward the center of the earth. It was a natural explanation for a large number of phenomena, but it could not easily be fitted to the observed behavior of a projectile. When released from the hand or from a sling, a stone does not drop straight to earth. Rather it continues to move in the direction toward which it was initially impelled even after its contact with the initial projector (hand or sling) is broken. Aristotle, who was a shrewd observer, knew how a projectile moves, and he patched his theory by conceiving the disturbed air as the source of a push which prolongs the projectile's motion after contact with the first propellant is broken. He does not seem to have been too pleased with this solution, for he provided at least two incompatible versions of it, and he was always a bit argumentative on this point. But for him it was never important; his fundamental interests lay elsewhere; he treated the projectile only as an aside and apparently only because it might create difficulty for his theory.

It did create difficulties, apparently almost immediately. John Philoponus, the sixth-century Christian commentator who records the earliest extant rejection of Aristotle's theory, attributes his own partial impetus-theory solution to the Hellenistic astronomer Hipparchus. Most other commentators were at least troubled by this aspect of Aristotle's thought. Perhaps no one, including its author, ever took the air as pusher quite seriously. But it was not until the fourteenth century, when difficulties in Aristotle's text were problems in their own right, that the issue of the projectile was fully faced and resolved by a substantial modification of Aristotle's theory. Though its source was a terrestrial problem, that modification proved to have immediate implications for astronomy.

Both the problem and its medieval solution can be recovered in brilliant detail from the *Questions on the Eight Books of Aristotle's Physics* (a typical title for scholastic science) by Oresme's teacher, Jean Buridan:

It is sought whether a projectile after leaving the hand of the projector is moved by the air or by what it is moved. . . . This question I judge to be very difficult because Aristotle, as it seems to me, has not solved it well. For he . . . [at one point] holds that the projectile swiftly leaves the place in which it was, and nature, not permitting a vacuum, rapidly sends air in behind to fill up the vacuum. The air moved in this way impinges upon the projectile and impels it along. This is repeated continually for a certain distance. . . . [But] it seems to me that many experiences show this method of proceeding to be valueless. . . .

[For example, one among many that Buridan gives,] a lance having a conical posterior as sharp as its anterior would be moved after projection just as swiftly as it would be without a sharp conical posterior. But surely the air following could not push on a sharp end in this way since the air would be easily divided by the sharpness [whereas it could push on a blunt end, thus moving the lance with the blunt end farther]. . . .

Thus we can and ought to say that in the stone or other projectile there is impressed something which is the motive force of that projectile. This is evidently better than falling back on the statement that the air would [continue to] move the projectile, for the air appears to resist. . . . [The projector] impresses a certain impetus or motive force into the moving body, which impetus acts in the direction toward which the mover moved the moving body, either up or down, or laterally or circularly. And by the amount the motor moves that moving body more swiftly, by the same amount it will impress in it a stronger impetus. It is by that impetus that the stone is moved after the projector ceases to move. But that impetus is continually decreased by the resisting air and by the gravity of the stone which inclines it in a direction contrary to that in which the impetus was naturally predisposed to move it. Thus the movement of the stone continually becomes slower until the impetus is so diminished or corrupted that the gravity of the stone wins out over it and moves the stone down to its natural place.[11]

This is only a fraction of Buridan's elaborate discussion, and countless parallel treatments can be found in the works of his successors. By the end of the fourteenth century impetus dynamics, in one of a number of forms very like Buridan's, had replaced Aristotelian dynamics in the work of the principal medieval scientists. The tradition endured: it was taught at Padua around the time Copernicus studied there; Galileo learned it from his master Bonamico at Pisa. Both of them used it, explicitly or implicitly, as did their contemporaries and successors. On a number of occasions and in a variety of ways the impetus theory played an essential role in the Copernican Revolution.

One of these roles we have already seen, though without recognizing it. Oresme's refutation of Aristotle's central argument for the earth's immobility takes the impetus theory, or something quite like it, for granted. On the Aristotelian theory of motion a vertically thrown stone must move along a radius fixed in space. If the earth moves while the stone is in the air, the stone (or arrow) cannot accompany it and will therefore not return to its point of departure. But it the earth's eastward motion endows the stone with an eastward impetus while the stone is still in contact with the projector, that impetus will endure and will cause the stone to pursue the moving earth even after contact is broken. The impetus theory enables the moving earth to endow terrestrial bodies with an internal propellant, and that propellant enables them to follow the earth afterward. Like his master Buridan, Oresme believed in the impetus theory, and though his refutation of Aristotle does not mention the theory explicitly, the refutation makes no sense without it. In one way or another the impetus theory is implicated in most of the arguments, both medieval and Renaissance, that make it possible to move the earth without leaving terrestrial bodies behind.

Some adherents of the impetus theory immediately extended it from the earth to the heavens. In the process they took a second long step toward the Copernicanism that was to come. Buridan himself said, in a passage that closely follows the preceding excerpt from his *Questions*:

Also, since the Bible does not state that appropriate [angelic] intelligences move the celestial bodies, it could be said that it does not appear necessary to posit intelligences of this kind. For it could [equally well] be answered that God, when He created the world, moved each of the celestial orbs as He pleased, and in moving them He impressed in them impetus which moved them without His having to move them any more except by the method of general influence whereby he concurs as a coagent in all things which take place. Thus on the seventh day He rested from all work which he had executed by committing to others the actions and the passions in turn. And these impetuses which He impressed in the celestial bodies were not decreased nor corrupted afterwards, because there was no inclination of the celestial bodies for other movements. Nor was there resistance which would be corruptive or repressive of that impetus.[12]

In Buridan's writings, perhaps for the first time, the heavens and the earth were at least tentatively subjected to a single set of laws, and the same suggestion was carried further by Buridan's pupil, Oresme. He suggested that, "when God created [the heavens] . . . , He impressed them with a certain quality and force of motion, just as He impressed terrestrial things with weight . . . ; it is just the same as a man building a clock and leaving it to run itself. Thus God left the heavens to be moved continually . . . according to the order [He had] established."[13] But to conceive the heavens as a terrestrial mechanism, a piece of clockwork, is to break the absolute dichotomy between the superlunary and sublunary regions. Though the impetus theorists followed the suggestion no further, at least during the Middle Ages, it was just this dichotomy, drawn from both Aristotle and theology, that had to be broken if the earth was to be made a planet.

The possibility of the earth's motion and the partial unification of terrestrial and celestial law were the impetus theory's two most direct contributions to the Copernican Revolution. Its most important contribution, however, was an indirect one, to which we shall revert briefly in the last chapter. Through its role in the evolution of Newtonian dynamics, the impetus theory helped to bring the Revolution to a successful close more than a century after Copernicus' death. Copernicus in the sixteenth century provided only a new mathematical description of the way the planets move; he was not successful in explaining why the planets moved as he said they did. Initially his mathematical astronomy made no physical sense, and it therefore posed new sorts of problems for his successors. Those problems were only resolved by Newton, whose dynamics supplied the missing keystone to Copernicus' mathematical system, and Newton's dynamics, even more than Copernicus' astronomy, depended on the prior scholastic analyses of motion.

Impetus dynamics is not Newtonian dynamics, but by pointing to new problems, new variables, and new abstractions impetus dynamics helped to pave the way for Newton's work. Before the impetus theory both Aristotle and experiment had testified that only rest endures. Buridan and some other impetus theorists declared that, unless resisted, motion too would endure forever, and they thus took a long step toward what we now know as Newton's First Law of Motion. Again, in a passage omitted from the descriptive quotation above, Buridan equated the quantity of impetus in a moving body with the product of the body's speed and its quantity of matter. The concept of impetus became very like, though not identical with, the modern concept of momentum; in Galileo's work the words "impetus" and "momentum" are often used interchangeably. Elsewhere, to give one final example, Buridan's discussion came very close to saying that the gravity (or weight) of a freely falling body impresses equal increments of impetus (and therefore of velocity) upon the body in equal intervals of time. Galileo was not the first of Buridan's successors to say precisely this and to derive from it, with the aid of other analytical devices supplied by the scholastics, the modern quantitative relation between time of fall and distance. Contributions like these give scholastic science an important role in the evolution of Newtonian dynamics, and Newtonian dynamics was the keystone in the structure of the new universe created by Copernicus and his successors.

During the seventeenth century, just when its full utility was being demonstrated for the first time, scholastic science was bitterly attacked by men trying to weave a radically new fabric of thought. The scholastics proved easy to ridicule, and the image has stuck. Medieval scientists more often found their problems in texts than in nature; many of those problems do not now seem problems at all; by modern standards the practice of science during the Middle Ages was incredibly inefficient. But how else could science have been reborn in the West? The centuries of scholasticism are the centuries in which the tradition of ancient science and philosophy was simultaneously reconstituted, assimilated, and tested for adequacy. As weak spots were discovered, they immediately became foci for the first effective research in the modern world. The great new scientific theories of the sixteenth and seventeenth centuries all originate from rents torn by scholastic criticism in the fabric of Aristotelian thought. Most of those theories also embody key concepts created by scholastic science. And more important even than these is the attitude that modern scientists inherited from their medieval predecessors: an unbounded faith in the power of human reason to solve the problems of nature. As the late Professor Whitehead remarked, "Faith in the possibility of science, generated antecedently to the development of modern scientific theory, is an unconscious derivative from medieval theology." [14]

Astronomy in the Age of Copernicus

In discussing late-medieval modifications of the Aristotelian-Ptolemaic tradition, we have said almost nothing about developed planetary astronomy. In fact there was almost none in Europe during the Middle Ages, partly because of the intrinsic difficulty of the mathematical texts and partly because the problem of the planets seemed so esoteric. Aristotle's *On the Heavens* described the entire universe in relatively simple terms; Ptolemy's elaborate *Almagest* dealt, for the most part, only with the computation of planetary position. Therefore, though the works of both Aristotle and Ptolemy were translated by the end of the twelfth century, Aristotelian logic, philosophy, and cosmology were assimilated far more rapidly than developed Ptolemaic astronomy. Thirteenth-century metaphysics rivals Aristotle's in profundity; fourteenth-century physics and cosmology exceed Aristotle's in depth and logical coherence. But Europeans produced no indigenous astronomical tradition to rival Ptolemy's until the middle of the fifteenth century, if then. The first widely known European astronomical treatise, written around 1233 by John of Holywood, slavishly copied an elementary Arabic treatise and devoted only one chapter to the planets compared with Ptolemy's nine. The next two centuries produced only commentaries on Holywood's book and a few unsuccessful rivals. Until two decades before Copernicus' birth in 1473 there was little concrete evidence of technically proficient planetary astronomy. Then it appeared in works like those of the German Georg Peuerbach (1423–1461) and his pupil Johannes Müller (1436–1476).

To Europeans of Copernicus' generation planetary astronomy was, therefore, almost a new field, and it was practiced in an intellectual and social environment quite different from any in which astronomy had been practiced before. In part that difference arose from the theological accretions to the astronomical tradition that we have examined in the work of Aquinas and Dante. Even more essential changes were produced by the logical and cosmological criticism of men like Buridan and Oresme. But these were medieval contributions, and Copernicus did not live during the Middle Ages. His lifetime, 1473–1543, occupied the central decades of the Renaissance and Reformation, and novelties characteristic of this later age were also effective in inaugurating and shaping his work.

Since stereotypes are most readily discarded during periods of general ferment, the turbulence of Europe during the Renaissance and Reformation itself facilitated Copernicus' astronomical innovation. Change in one field decreases the hold of stereotypes in others. Radical innovations in science have repeatedly occurred during periods of national or international convulsion, and Copernicus' lifetime was such a period. The Moslems again threatened to absorb vast areas of a Europe now convulsed by the dynastic rivalries through which the nation-state replaced the feudal monarchy. A new commercial aristocracy, accompanied by rapid changes in economic institutions and in technology, began to rival the older aristocracies of Church and landed nobility. Luther and Calvin led the first successful revolts against Catholic religious hegemony. In an age marked by such obvious upheavals in political, social, and religious life, an innovation in planetary astronomy could at first seem no innovation at all.

Specific characteristics of the age had more concrete effects on astronomy. The Renaissance was, for example, a period of voyages and explorations. Fifty years before Copernicus' birth Portuguese voyages along the African coast had begun to excite the imagination and avarice of Europeans. Columbus' first landfall in America, made when Copernicus was nineteen, only capped this earlier series of explorations and provided the basis for a new series. Successful voyages demanded improved maps and navigational techniques, and these depended in part on increased knowledge of the heavens. Prince Henry the Navigator, the organizer and director of the early Portuguese voyages, constructed one of the earliest European observatories. Exploration therefore helped to create a demand for expert European astronomers, and having done so, it partially changed their attitude toward their field. Each new voyage disclosed new territory, new products, and new people. Men rapidly learned how wrong ancient descriptions of the earth could be. In particular, they learned how wrong Ptolemy could be, for Ptolemy had been the greatest geographer as well as the greatest astronomer and astrologer of antiquity. The astronomer's awareness — an awareness which we shall shortly discover in Copernicus himself — that Renaissance man could at last correct Ptolemy's geography prepared him for changes in his own closely related field.

Agitation for calendar reform had an even more direct and dramatic effect on the practice of Renaissance astronomy, for the study of calendars brought the astronomer face to face with the inadequacy of existing computational techniques. The cumulative errors of the Julian calendar had been recognized much earlier, and proposals for calendar reform date from the thirteenth century or before. But these proposals were ineffective until the sixteenth century, when the increasing size of political, economic, and administrative units placed a new premium upon an efficient and uniform means of dating. Then reform became an official Church project, with results for astronomy that are well illustrated in the biography of Copernicus himself. Early in the sixteenth century Copernicus was asked to advise the papacy about calendar reform. He declined and urged that reform be postponed, because he felt that existing astronomical observations and theories did not yet permit the design of a truly adequate calendar. When Copernicus listed the aspects of contemporary astronomy that had led him to consider his radical theory, he began, "For, first, the mathematicians are so unsure of the movements of the Sun and Moon that they cannot even explain or observe the constant length of the seasonal year" (see p. 137 below). Reform of the calendar demanded, said Copernicus, reform in astronomy. The preface of his *De Revolutionibus* closed with the suggestion that his new theory might make a new calendar possible. The Gregorian calendar, first adopted in 1582, was in fact based upon computations that made use of Copernicus' work.

Recognition of the inadequacies in existing techniques for astronomical computation was heightened by still another aspect of Renaissance life. During the fifteenth century, Europe experienced a second great intellectual revival associated with a second rediscovery of classical models. Unlike its twelfth-century predecessor, the Renaissance revival of learning was not primarily a scientific revival. Most of the newly recovered documents exemplified ancient literature, art, and architecture, subjects whose great tradition was then little known in the West principally because Islamic culture had been indifferent to it. The manuscripts discovered in the fifteenth century did, however, include a few important works of Hellenistic mathematics and, more important, a great many authentic Greek versions of scientific classics previously known only in Arabic. As a result, the Ptolemaic system's recognized failure correctly to predict celestial motions could no longer be blamed upon errors accumulated in transmission and translation. Astronomers could no longer believe that astronomy had declined since Ptolemy.

Peuerbach, for example, began his career in astronomy by working from second-hand translations of the *Almagest* transmitted via Islam. From them he was able to reconstruct a more adequate and complete account of Ptolemy's system than any known before. But his work only convinced him that a truly adequate astronomy could not be derived from Arabic sources. Astronomers, he felt, would have to work from Greek originals, and he was about to depart for Italy to examine manuscripts available there when he died in 1461. His successors, particularly Johannes Müller, did work from the Greek versions, and they then discovered that even Ptolemy's original formulation was inadequate. By making available sound texts of ancient authors, fifteenth-century scholars helped Copernicus' immediate predecessors to recognize that it was time for a change.

Developments like those discussed above can help us understand why the Copernican Revolution occurred when it did. They are essential parts of the climate for astronomical upheaval. But there are other, more intellectual, aspects of the Renaissance which played a somewhat different role in the Revolution. They are associated with humanism, the dominant learned movement of the age, and their effect was less upon the Revolution's timing than upon its shape. Humanism was not principally a scientific movement. The humanists themselves were often bitterly opposed to Aristotle, the scholastics, and the entire tradition of university learning. Their sources were the newly recovered literary classics, and, like literary men in other ages, many of them rejected the scientific enterprise as a whole. The early humanist poet Petrarch sounds a typical note, strangely and significantly reminiscent of Augustine's earlier depreciation of science. "Even if all these things were true, they help in no way toward a happy life, for what does it advantage us to be familiar with the nature of animals, birds, fishes, and reptiles, while we are ignorant of the nature of the race of man to which we belong, and do not know or care whence we come or whither we go." [15] If humanism had been the only intellectual movement of the Renaissance, the Copernican Revolution might have been long postponed. The work of Copernicus and his astronomical contemporaries belongs squarely in that university tradition which the humanists most ridiculed.

The humanists did not, however, succeed in stopping science. During the Renaissance a dominant humanistic tradition outside the universities existed side by side with a continuing scientific tradition within university walls. In consequence the first scientific effect of the humanists' dogmatic anti-Aristotelianism was to facilitate for others a break with the root concepts of Aristotle's science. A second but more important effect was the surprising fertilization of science by the strong otherworldly strain that characterized humanist thought. From this aspect of humanism, a first hint of which is contained in the preceding quotation from Petrarch, some Renaissance scientists, like Copernicus, Galileo, and Kepler, seem to have drawn two decidedly un-Aristotelian ideas: a new belief in the possibility and importance of discovering simple arithmetic and geometric regularities in nature, and a new view of the sun as the source of all vital principles and forces in the universe.

The otherworldliness of humanism derived from a well-defined philosophical tradition which had greatly influenced Augustine and other early Church fathers but which had been temporarily eclipsed by the twelfth-century recovery of Aristotle's writings. That tradition, unlike the Aristotelian, discovered reality in a changeless world of spirit rather than in the transient affairs of everday life. Plato, who is the tradition's ultimate source, often seems to dismiss the objects of this world as mere imperfect shadows of an eternal world of ideal objects or "forms" existing outside of space and time. His followers, the so-called Neoplatonists emphasized this tendency in their master's thought to the exclusion of all others. Their mystical philosophy, which many humanists took as a model, recognized only a transcendent reality. Yet for all its mysticism, Neoplatonism contained elements that gave a significant new direction to the science of the Renaissance.

The Neoplatonist leaped at once from the changeable and corruptible world of everyday life to the eternal world of pure spirit, and mathematics showed him how to make the leap. For him mathematics exemplified the eternal and real amid the imperfect and fluctuating appearances of the terrestrial world. The triangles and circles of plane geometry were the archetypes of all Platonic forms. They existed nowhere — no line or point drawn on paper satisfies Euclid's postulates — but they were endowed with certain eternal and necessary properties which the mind alone could discover and which, once discovered, could be observed dimly mirrored in the objects of the real world. The Pythagoreans, who had also envisioned the real world as a shadow of the eternal world of mathematics, exemplified the ideal of terrestrial science in their discovery that uniform strings whose lengths are in the simple numerical ratios $1 : \frac{3}{4} : \frac{2}{3} : \frac{1}{2}$ produce harmonious sounds. The mathematical strain in Neoplatonic thought is often attributed to Pythagoras and identified as Neopythagoreanism.

Plato himself emphasized the necessity of mathematics as training for the mind in pursuit of forms; over the door of his Academy he is said to have inscribed, "Let no one destitute of geometry enter my doors." [16] The Neoplatonists went further. They found in mathematics the key to the essential nature of God, the soul, and the world soul which was the universe. A typical passage from the fifth-century Neoplatonist Proclus perfectly displays a part of this mystical vision of mathematics:

> The soul [of the world], therefore, is by no means to be compared to a smooth tablet, void of all reasons; but she is an ever-written tablet, herself inscribing the characters in herself, of which she derives an eternal plenitude from intellect. . . . All mathematical species, therefore, have a primary subsistence in the soul: so that, before sensible numbers, there are to be found in her inmost recesses, self-moving numbers; vital figures, prior to the apparent; ideal proportions of harmony previous to concordant sounds; and invisible orbs, prior to the bodies which revolve in a circle. . . . [We] must conceive all these as subsisting ever vitally, and intellectually, as the exemplars of apparent numbers, figures, reasons and motions. And here we must follow the doctrine of Timaeus, who derives the origin, and consummates the fabric of the soul, from mathematical forms, and reposes in her nature the causes of everything which exists.[17]

Proclus and the humanists who espoused his cause are a long way from the physical sciences. But they occasionally influenced their more scientifically inclined contemporaries, and the result was a new search by many late Renaissance scientists for simple geometric and arithmetic regularities in nature. Copernicus' friend and teacher at Bologna, Domenico Maria de Novara, was a close associate of the Florentine Neoplatonists who translated Proclus and other authors of his school. Novara himself was among the first to criticize the Ptolemaic planetary theory on Neoplatonic grounds, believing that no system so complex and cumbersome could represent the true mathematical order of nature. When Novara's pupil Copernicus complained that the Ptolemaic astronomers "seem to violate the first principle of uniformity in motion" and that they have been unable to "deduce the principal thing — namely the shape of the Universe and the unchangeable symmetry of its parts" (see p. 138 below), he was participating in the same Neoplatonic tradition. The Neoplatonic strain shows even more strongly in Copernicus' great successor, Kepler. As we shall see, the search for simple numerical relations runs through and motivates most of Kepler's work.

The origin of the liaison between Neoplatonism and sunworship is more obscure, but a hint about the ties that bound them together can be found in the preceding quotation from Proclus. Neoplatonic thought could never quite dispense with the real world. The "vital figures" and "invisible orbs" that Proclus found in the soul of the world, or in God, might be the primary philosophical entities, the only things that have complete reality and existence. But the Neoplatonist could not avoid granting some sort of existence to the imperfect bodies attested by his senses. He therefore regarded them as second-rate copies generated by the "vital figures" themselves. As Proclus said, the mathematical forms that determine the nature of the world soul were also the "causes of everything which exists." They generated countless debased and materialized copies of their own purely intellectual substance. The Neoplatonist's God was a self-duplicating procreative principle whose immense potency was demonstrated by the very multiplicity of the forms that emanated from Him. In the material universe this fecund Deity was suitably represented by the sun whose visible and invisible emanations gave light, warmth, and fertility to the universe.

This symbolic identification of the sun and God is found repeatedly in Renaissance literature and art. Marsilio Ficino, a central figure in the fifteenth-century humanist and Neoplatonic academy of Florence, gave it typical expression in his treatise *On the Sun*:

Nothing reveals the nature of the Good [which is God] more fully than the light [of the sun]. First, light is the most brilliant and clearest of sensible objects. Second, there is nothing which spreads out so easily, broadly, or rapidly as light. Third, like a caress, it penetrates all things harmlessly and most gently. Fourth, the heat which accompanies it fosters and nourishes all things and is the universal generator and mover. . . . Similarly the Good is itself spread everywhere, and it soothes and entices all things. It does not work by compulsion, but through the love which accompanies it, like heat [which accompanies light]. This love allures all objects so that they freely embrace the Good. . . . Perhaps light is itself the celestial spirit's sense of sight, or its act of seeing, operating from a distance, linking all things to heaven, yet never leaving heaven nor mixing with external things. . . . Just look at the skies, I pray you, citizens of heavenly fatherland. . . . The sun can signify God Himself to you, and who shall dare to say the sun is false.[18]

With Ficino as with Proclus, we are a long way from science. Ficino does not seem to understand astronomy. He certainly made no attempt to reconstruct it. Though the sun has a new significance in Ficino's universe, it retains its old position. Yet that position was no longer appropriate. Ficino wrote, for example, that the sun was created first and in the center of the heavens. Surely no lesser position in space or in time could be compatible with the sun's dignity and creative function. But the position was not compatible with Ptolemaic astronomy, and the resulting difficulties for Neoplatonism may have helped Copernicus to conceive a new system constructed about a central sun. In any case they gave him an argument for the new system. As soon as he had discussed the new position of the sun, Copernicus adverted to the fitness of his new cosmology (see p. 179 below). His authorities are immediately Neoplatonic:

> In the middle of all sits Sun enthroned. In this most beautiful temple could we place this luminary in any better position from which he can illuminate the whole at once? He is rightly called the Lamp, the Mind, the Ruler of the Universe; Hermes Trismegistus names him the Visible God, Sophocles' Electra calls him the All-seeing. So the sun Sits as upon a royal throne ruling his children the planets which circle round him.

Neoplatonism is explicit in Copernicus' attitude toward both the sun and mathematical simplicity. It is an essential element in the intellectual climate that gave birth to his vision of the universe. But it is often hard to tell whether any given Neoplatonic attitude is posterior or antecedent to the invention of his new astronomy in Copernicus' thought. No similar ambiguity exists with respect to the later Copernicans. Kepler, for example, the man who made the Copernican system work, is quite explicit about his reasons for preferring Copernicus' proposal, and among them is the following:

> [The sun] is a fountain of light, rich in fruitful heat, most fair, limpid, and pure to the sight, the source of vision, portrayer of all colors, though himself empty of color, called king of the planets for his motion, heart of the world for his power, its eye for his beauty, and which alone we should judge worthy of the Most High God, should he be pleased with a material domicile and choose a place in which to dwell with the blessed angels. . . . For if the Germans elect him as Caesar who has most power in the whole empire, who would hesitate to confer the votes of the celestial motions on him who already has been administering all other movements and changes by the benefit of the light which is entirely his possession? . . . [Hence] by the highest right we return to the sun, who alone appears, by virtue of his dignity and power, suited for this motive duty and worthy to become the home of God himself, not to say the first mover.[19]

Until after Copernicus' death the mathematical magic and the sun worship that are so marked in Kepler's research remained the principal points of explicit contact between Renaissance Neoplatonism and the new astronomy. But late in the sixteenth century a third aspect of Neoplatonism merged with Copernicanism and helped to reshape the structure of Copernicus' universe. Unlike the Deity of the Neoplatonists, whose perfection was measured by his immense fecundity, the God of Aquinas and Aristotle had been conceived as an architect who displayed His perfection in the neatness and order of His creation. Aquinas' God was well suited to the finite Aristotelian cosmos, but the God of the Neoplatonists was not so easily bounded. If God's perfection is measured by the extent and multiplicity of his procreation, a larger and more populous universe must connote a more perfect Deity. To many Neoplatonists the finitude of Aristotle's universe was therefore incompatible with God's perfection. His infinite goodness would, they felt, be satisfied only by an infinite act of creation. Even before Copernicus the resulting vision of a multipopulated universe, infinite in extent, had been the source of important divergences from Aristotelian doctrine. During the Renaissance the revived emphasis on God's infinite creativity may have been a significant element in the climate of opinion that bred Copernicus' innovation. Certainly, as we shall see later, it was a major factor in the post-Renaissance transition from Copernicus' finite universe to the infinite space of the Newtonian world machine.

Neoplatonism completes the conceptual stage setting for the Copernican Revolution, at least as we shall examine it here. For an astronomical revolution it is a puzzling stage, because it is set with so few astronomical properties. Their absence, however, is just what makes the setting important. Innovations in a science need not be responses to novelties within that science at all. No fundamental astronomical discovery, no new sort of astronomical observation, persuaded Copernicus of ancient astronomy's inadequacy or of the necessity for change. Until half a century after Copernicus' death no potentially revolutionary changes occurred in the data available to astronomers. Any possible understanding of the Revolution's timing and of the factors that called it forth must, therefore, be sought principally outside of astronomy, within the larger intellectual milieu inhabited by astronomy's practitioners. As we suggested at the beginning of this chapter,

Copernicus began his cosmological and astronomical researches very nearly where Aristotle and Ptolemy had stopped. In that sense he is the immediate heir of the ancient scientific tradition. But his inheritance took almost two milleniums to reach him. In the interim the very process of rediscovery, the medieval integration of science and theology, the centuries of scholastic criticism, and the new currents of Renaissance life and thought, all had combined to change men's attitude towards the scientific heritage that they learned in school. Just how great, and yet how strangely small, this essential change could be we shall discover in the next chapter as we take up Copernicus' innovation.

5

COPERNICUS' INNOVATION

Copernicus and the Revolution

The publication of Copernicus' *De Revolutionibus Orbium Caelestium* in 1543 inaugurates the upheaval in astronomical and cosmological thought that we call the Copernican Revolution. To this point we have dealt only with the background of that Revolution, setting the stage upon which the Revolution occurred. Now we turn to the Revolution itself, dealing first, in this chapter, with Copernicus' contributions to it. So far as possible we shall discover those contributions in Copernicus' own words, drawn from the *De Revolutionibus*, the book that presented the new astronomy to the world. Almost immediately we shall encounter difficulties and incongruities upon whose resolution depends our understanding of the Copernican Revolution or, since that Revolution is in many respects typical, of any other major conceptual upheaval in the sciences.

The *De Revolutionibus* is for us a problem text. Some of its problems derive simply from the intrinsic difficulties of its subject matter. All but the introductory First Book is too mathematical to be read with understanding by anyone except a technically proficient astronomer. We must deal with its essential technical contributions in relatively nonmathematical paraphrase, much like that employed in treating the *Almagest*, and we shall by-pass in this process certain of the essential problems that the *De Revolutionibus* presented to its sixteenth-century readers. Had Copernicus propounded the new astronomy in the simplified form to which we shall frequently resort in this chapter, its reception might have been quite different. Opposition to a more comprehensible work might, for example, have been marshaled sooner. Our first problem is therefore the barrier which a lack of technical proficiency places between us and the central books of the work that inaugurated the Revolution.

But the technical obscurity of the *De Revolutionibus*, though it must be recognized at the start, is neither the most difficult nor the most important sort of problem inherent in Copernicus' work. The principal difficulties of the *De Revolutionibus* and the ones that we may not evade arise rather from the apparent incompatibility between that text and its role in the development of astronomy. In its consequences the *De Revolutionibus* is undoubtedly a revolutionary work. From it derive a fundamentally new approach to planetary astronomy, the first accurate and simple solution of the problem of the planets, and ultimately, with other fibers added to the pattern, a new cosmology. But, to any reader aware of this outcome, the *De Revolutionibus* itself must be a constant puzzle and paradox, for, measured in terms of its consequences, it is a relatively staid, sober, and unrevolutionary work. Most of the essential elements by which we know the Copernican Revolution — easy and accurate computations of planetary position, the abolition of epicycles and eccentrics, the dissolution of the spheres, the sun a star, the infinite expansion of the universe — these and many others are not to be found anywhere in Copernicus' work. In every respect except the earth's motion the *De Revolutionibus* seems more closely akin to the works of ancient and medieval astronomers and cosmologists than to the writings of the succeeding generations who based their work upon Copernicus' and who made explicit the radical consequences that even its author had not seen in his work.

The significance of the *De Revolutionibus* lies, then, less in what it says itself than in what it caused others to say. The book gave rise to a revolution that it had scarcely enunciated. It is a revolution-making rather than a revolutionary text. Such texts are a relatively frequent and extremely significant phenomenon in the development of scientific thought. They may be described as texts that shift the direction in which scientific thought develops; a revolution-making work is at once the culmination of a past tradition and the source of a novel future tradition. As a whole the *De Revolutionibus* stands almost entirely within an ancient astronomical and cosmological tradition; yet within its generally classical framework are to be found a few novelties which shifted the direction of scientific thought in ways unforeseen by its author and which gave rise to a rapid and complete break with the ancient tradition. Viewed in a perspective provided by the history of astronomy, the *De Revolutionibus* has a dual nature. It is at once ancient and modern, conservative and radical. Therefore its significance can be discovered only by looking simultaneously to its past and to its future, to the tradition from which it derived and to the tradition which derives from it.

That double view of a single work is the principal problem of this chapter. What is the relation of Copernicus to the ancient astronomical tradition within which he was educated? More precisely, what aspects of that tradition led him to believe that some astronomical innovation was essential, that certain aspects of ancient cosmology and astronomy must be rejected? And, having resolved to break with an old tradition, to what extent was he still necessarily bound by it as the only source of those intellectual and observational tools required for the practice of astronomy? Again, what is Copernicus' relation to the tradition of modern planetary astronomy and cosmology? Given the limitations imposed by the training and tools of classical astronomy, what creative innovations could his work contain? How could those innovations, which ultimately produced a radically new astronomy and cosmology, be embedded initially in a predominantly classical frame? And how could those novelties be recognized and adopted by his successors? These problems and their corollaries are symptomatic of the real difficulties of the *De Revolutionibus* or of any scientific work which, though born within one tradition of scientific thought, is the source of a new tradition that ultimately destroys its parent.

Motives for Innovation — Copernicus' Preface

Copernicus is among that small group of Europeans who first revived the full Hellenistic tradition of technical mathematical astronomy which in antiquity had culminated in the work of Ptolemy. The *De Revolutionibus* was modeled on the *Almagest*, and it was directed almost exclusively to that small group of contemporary astronomers equipped to read Ptolemy's treatise. With Copernicus we return for the first time to the sort of technical astronomical problem with which we last dealt in Chapter 3 when examining the developed Ptolemaic system. In fact we return to the same problem. The *De Revolutionibus* was written to solve the problem of the planets, which, Copernicus felt, Ptolemy and his successors had left unsolved. In

Copernicus' work the revolutionary conception of the earth's motion is initially an anomalous by-product of a proficient and devoted astronomer's attempt to reform the techniques employed in computing planetary position. That is the first significant incongruity of the *De Revolutionibus*, the disproportion between the objective that motivated Copernicus' innovation and the innovation itself. It can be discovered almost at the start of the prefatory letter that Copernicus prefixed to the *De Revolutionibus* in order to sketch the motive, the source, and the nature of his scientific achievement.[1]

TO THE MOST HOLY LORD, POPE PAUL III

The Preface of Nicholas Copernicus to the
Books of the Revolutions

I may well presume, most Holy Father, that certain people, as soon as they hear that in this book about the Revolutions of the Spheres of the Universe I ascribe movement to the earthly globe, will cry out that, holding such views, I should at once be hissed off the stage. For I am not so pleased with my own work that I should fail duly to weigh the judgment which others may pass thereon; and though I know that the speculations of a philosopher are far removed from the judgment of the multitude — for his aim is to seek truth in all things as far as God has permitted human reason so to do — yet I hold that opinions which are quite erroneous should be avoided.

Thinking therefore within myself that to ascribe movement to the Earth must indeed seem an absurd performance on my part to those who know that many centuries have consented to the establishment of the contrary judgment, namely that the Earth is placed immovably as the central point in the middle of the Universe, I hesitated long whether, on the one hand, I should give to the light these my Commentaries written to prove the Earth's motion, or whether, on the other hand, it were better to follow the example of the Pythagoreans and others who were wont to impart their philosophic mysteries only to intimates and friends, and then not in writing but by word of mouth, as the letter of Lysis to Hipparchus witnesses. [This letter, which Copernicus had at one time intended to include in the *De Revolutionibus*, describes the Pythagorean and Neoplatonic injunction against revealing nature's secrets to those who are not initiates of a mystical cult. Reference to it here exemplifies Copernicus' participation in the Renaissance revival of Neoplatonism discussed in the last chapter.] In my judgment they did so not, as some would have it, through jealousy of sharing their doctrines, but as fearing lest these so noble and hardly won discoveries of the learned should be despised by such as either care not to study aught save for gain, or — if by the encouragement and example of others they are stimulated to philosophic liberal pursuits — yet by reason of the dullness of their wits are in the company of philosophers as drones among bees. Reflecting thus, the thought of the scorn which I had to fear on account of the novelty and incongruity of my theory, well-nigh induced me to abandon my project.

These misgivings and actual protests have been overcome by my friends . . . [one of whom] often urged and even importuned me to publish this work which I had kept in store not for nine years only, but to a fourth period of nine years. . . . They urged that I should not, on account of my fears, refuse any longer to contribute the fruits of my labors to the common advantage of those interested in mathematics. They insisted that, though my theory of the Earth's movement might at first seem strange, yet it would appear admirable and acceptable when the publication of my elucidatory comments should dispel the mists of paradox. Yielding then to their persuasion I at last permitted my friends to publish that work which they have so long demanded.

That I allow the publication of these my studies may surprise your Holiness the less in that, having been at such travail to attain them, I had already not scrupled to commit to writing my thoughts upon the motion of the Earth. [Some years before the publication of the *De Revolutionibus* Copernicus had circulated among his friends a short manuscript called the *Commentariolus*, describing an earlier version of his sun-centered astronomy. A second advance report of Copernicus' major work, the *Narratio Prima* by Copernicus' student, Rheticus, had appeared in 1540 and again in 1541.] How I came to dare to conceive such motion of the Earth, contrary to the received opinion of the Mathematicians and indeed contrary to the impression of the senses, is what your Holiness will rather expect to hear. So I should like your Holiness to know that I was induced to think of a method of computing the motions of the spheres by nothing else than the knowledge that the Mathematicians are inconsistent in these investigations.

For, first, the mathematicians are so unsure of the movements of the Sun and Moon that they cannot even explain or observe the constant length of the seasonal year. Secondly, in determining the motions of these and of the other five planets, they use neither the same principles and hypotheses nor the same demonstrations of the apparent motions and revolutions. So some use only homocentric circles [the Aristotelian system, derived by Aristotle from Eudoxus and Callippus, and revived in Europe shortly before Copernicus' death by the Italian astronomers Fracastoro and Amici], while others [employ] eccentrics and epicycles. Yet even by these means they do not completely attain their ends. Those who have relied on homocentrics, though they have proven that some different motions can be compounded therefrom, have not thereby been able fully to establish a system which agrees with the phenomena. Those again who have devised eccentric systems, though they appear to have well-nigh established the seeming motions by calculations agreeable to their assumptions, have yet made many admissions [like the use of the equant] which seem to violate the first principle of uniformity in motion. Nor have they been able thereby to discern or deduce the principal thing — namely the shape of the Universe and the unchangeable symmetry of its parts. With them it is as though an artist were to gather the hands, feet, head and other members for his images from diverse models, each part excellently drawn, but not related to a single body, and since they in no way match each other, the result would be monster rather than man. So in the course of their exposition, which the mathematicians call their system, . . . we find that they have either omitted some indispensable detail or introduced something foreign and wholly irrelevant. This would of a surety not have been so had they followed fixed principles; for if their hypotheses were not misleading, all inferences based thereon might be surely verified. Though my present assertions are obscure, they will be made clear in due course.

An honest appraisal of contemporary astronomy, says Copernicus, shows that the earth-centered approach to the problem of the planets is hopeless. The traditional techniques of Ptolemaic astronomy have not and will not solve that problem; instead they have produced a monster; there must, he concludes, be a fundamental error in the basic concepts of traditional planetary astronomy. For the first time a technically competent astronomer had rejected the time-honored scientific tradition for reasons internal to his science, and this professional awareness of technical fallacy inaugurated the Copernican Revolution. A felt necessity was the mother of Copernicus' invention. But the feeling of necessity was a new one. The astronomical tradition had not previously seemed monstrous. By Copernicus' time a metamorphosis had occurred, and Copernicus' preface brilliantly describes the felt causes of that transformation.

Copernicus and his contemporaries inherited not only the *Almagest* but also the astronomies of many Islamic and a few European astronomers who had criticized and modified Ptolemy's system. These are the men to whom Copernicus refers as "the mathematicians." One had added or subtracted a few small circles; another had employed an epicycle to account for a planetary irregularity that Ptolemy had originally treated with an eccentric; still another had invented a means unknown to Ptolemy of accounting for small deviations from the motion predicted by a one-epicycle one-deferent system; others had, with new measurements, altered the rates at which the compounded circles of Ptolemy's system rotated. There was no longer one Ptolemaic system, but a dozen or more, and the number was multiplying rapidly with the multiplication of technically proficient astronomers. All these systems were modeled on the system of the *Almagest*, and all were therefore "Ptolemaic." But because there were so many variant systems, the adjective "Ptolemaic" had lost much of its meaning. The astronomical tradition had become diffuse; it no longer fully specified the techniques that an astronomer might employ in computing planetary position, and it could not therefore specify the results that he would obtain from his computations. Equivocations like these deprived the astronomical tradition of its principal source of internal strength.

Copernicus' monster has other faces. None of the "Ptolemaic" systems which Copernicus knew gave results that quite coincided with good naked-eye observations. They were no worse than Ptolemy's results, but they were also no better. After thirteen centuries of fruitless research a perceptive astronomer might well wonder, as Ptolemy could not have, whether further attempts within the same tradition could conceivably be successful. Besides, the centuries that had intervened between Ptolemy and Copernicus had magnified the errors of the traditional approach, thus providing an additional source of discontent. The motions of a system of epicycles and deferents are not unlike those of the hands of a clock, and the apparent error of a clock increases with the passage of time. If a clock loses, say, 1 second per decade, its error may not be apparent at the end of a year or the end of ten. But the error can scarcely be evaded after a millenium, when it will have increased to almost 2 minutes. Since Copernicus and his contemporaries possessed astronomical data extending over a time span thirteen centuries longer than that covered by Ptolemy's data, they could impose a far more sensitive check upon their systems. They were necessarily more aware of the errors inherent in the ancient approach.

The passage of time also presented the sixteenth-century astronomer with a counterfeit problem which ironically was even more effective than the real motion of the planets in fostering recognition of the errors in the Ptolemaic method. Many of the data inherited by Copernicus and his colleagues were bad data which placed the planets and stars in positions that they had never occupied. Some of the erroneous records had been collected by poor observers; others had once been based upon good observations but had been miscopied or misconstrued during the process of transmission. No simple planetary system — Ptolemy's, Copernicus', Kepler's, or Newton's — could have reduced to order the data that Renaissance astronomers thought they had to explain. The complexity of the problem presented by Renaissance data transcended that of the heavens themselves. Copernicus was himself a victim of the data that had originally aided him in rejecting the Ptolemaic system. His own system would have given far better results if he had been as skeptical about his predecessors' observations as he was about their mathematical systems.

Diffuseness and continued inaccuracy — these are the two principal characteristics of the monster described by Copernicus. In so far as the Copernican Revolution depended upon explicit changes within the astronomical tradition itself, these are its major sources. But they are not the only ones. We may also ask why Copernicus was able to recognize the monster. Some of the tradition's apparent metamorphosis must have been in the eye of the beholder, for the tradition had been diffuse and inaccurate before. In fact we have already considered this question. Copernicus' awareness of monstrosity depended upon that larger climate of philosophical and scientific opinion whose genesis and nature were described in the last chapter. From the state of contemporary astronomy a man without Copernicus' Neoplatonic bias might have concluded merely that the problem of the planets could have no solution that was simultaneously simple and precise. Similarly, an astronomer unacquainted with the tradition of scholastic criticism might have been unable to develop parallel criticisms for his own field. These and other novelties developed in the last chapter are main currents of Copernicus' time. Though he seems unaware of them, Copernicus was carried by these philosophical currents, as his contemporaries were unwittingly carried by the motion of the earth. Copernicus' work remains incomprehensible unless viewed in its relation to both the internal state of astronomy and the larger intellectual climate of the age. Both together produced the monster.

Discontent with a recognized monster was, however, only the first step toward the Copernican Revolution. Next came a search whose beginnings are described in the remaining portions of Copernicus' prefatory letter:

> I pondered long upon this uncertainty of mathematical tradition in establishing the motions of the system of the spheres. At last I began to chafe

that philosophers could by no means agree on any one certain theory of the mechanism of the Universe, wrought for us by a supremely good and orderly Creator, though in other respects they investigated with meticulous care the minutest points relating to its circles. [Note how Copernicus equates "orderly" with "mathematically neat," an aspect of his Neoplatonism from which any good Aristotelian would have vehemently dissented. There are other sorts of orderliness.] I therefore took pains to read again the works of all the philosophers on whom I could lay hand to seek out whether any of them had ever supposed that the motions of the spheres were other than those demanded by the mathematical schools. I found first in Cicero that Hicetas [of Syracuse, fifth century B.C.] had realized that the Earth moved. Afterwards I found in Plutarch that certain others had held the like opinion. I think fit here to add Plutarch's own words, to make them accessible to all:

"The rest hold the Earth to be stationary, but Philolaus the Pythagorean [fifth century B.C.] says that she moves around the [central] fire on an oblique circle like the Sun and Moon. Heraclides of Pontus and Ecphantus the Pythagorean [fourth century B.C.] also make the Earth to move, not indeed through space but by rotating round her own center as a wheel on an axle from West to East."

Taking advantage of this I too began to think of the mobility of the Earth; and though the opinion seemed absurd, yet knowing now that others before me had been granted freedom to imagine such circles as they chose to explain the phenomena of the stars, I considered that I also might easily be allowed to try whether, by assuming some motion of the Earth, sounder explanations than theirs for the revolution of the celestial spheres might so be discovered.

Thus assuming motions, which in my work I ascribe to the Earth, by long and frequent observations I have at last discovered that, if the motions of the rest of the planets be brought into relation with the circulation of the Earth and be reckoned in proportion to the circles of each planet, not only do their phenomena presently ensue, but the orders and magnitudes of all stars and spheres, nay the heavens themselves, become so bound together that nothing in any part thereof could be moved from its place without producing confusion of all the other parts and of the Universe as a whole. . . . [Copernicus here points to the single most striking difference between his system and Ptolemy's. In the Copernican system it is no longer possible to shrink or expand the orbit of any one planet at will, holding the others fixed. Observation for the first time can determine the order and the relative dimensions of all the planetary orbits without resort to the hypothesis of space-filling spheres. We shall discuss the point more fully when we compare Copernicus' system with Ptolemy's.]

I doubt not that gifted and learned mathematicians will agree with me if they are willing to comprehend and appreciate, not superficially but thoroughly, according to the demands of this science, such reasoning as I bring to bear in support of my judgment. But that learned and unlearned alike may see that I shrink not from any man's criticism, it is to your Holiness rather than anyone else that I have chosen to dedicate these studies of mine, since in this remote corner of Earth in which I live you are regarded as the most eminent by virtue alike of the dignity of your Office and of your love of letters and science. You by your influence and judgment can readily hold the slanderers from biting, though the proverb hath it that there is no remedy against a sycophant's tooth. It may fall out, too, that idle babblers, ignorant of mathematics, may claim a right to pronounce a judgment on my work, by reason of a certain passage of Scripture basely twisted to suit their purpose. Should any such venture to criticize and carp at my project, I make no account of them; I consider their judgment rash, and utterly despise it. I well know that even Lactantius, a writer in other ways distinguished but in no sense a mathematician, discourses in a most childish fashion touching the shape of the Earth, ridiculing even those who have stated the Earth to be a sphere. Thus my supporters need not be amazed if some people of like sort ridicule me too.

Mathematics are for mathematicians, and they, if I be not wholly deceived, will hold that these my labors contribute somewhat even to the Commonwealth of the Church, of which your Holiness is now Prince. For not long since, under Leo X, the question of correcting the ecclesiastical calendar was debated in the Council of the Lateran. It was left undecided for the sole cause that the lengths of the years and months and the motions of the Sun and Moon were not held to have been yet determined with sufficient exactness. From that time on I have given thought to their more accurate observation, by the advice of that eminent man Paul, Lord Bishop of Sempronia, sometime in charge of that business of the calendar. What results I have achieved therein, I leave to the judgment of learned mathematicians and of your Holiness in particular. And now, not to seem to promise your Holiness more than I can perform with regard to the usefulness of the work, I pass to my appointed task.

"Mathematics are for mathematicians." There is the first essential incongruity of the *De Revolutionibus*. Though few aspects of Western thought were long unaffected by the consequences of Copernicus' work, that work itself was narrowly technical and professional. It was mathematical planetary astronomy, not cosmology or philosophy, that Copernicus found monstrous, and it was the reform of mathematical astronomy that alone compelled him to move the earth. If his contemporaries were to follow him, they would have to learn to understand his detailed mathematical arguments about planetary position, and they would have to take these abstruse arguments more seriously than the first evidence of their senses. The Copernican Revolution was not primarily a revolution in the mathematical techniques employed to compute planetary position, but it began as one. In recognizing the need for and in developing these new techniques, Copernicus made his single original contribution to the Revolution that bears his name.

Copernicus was not the first to suggest the earth's motion, and he did not claim to have rediscovered the idea for himself. In his preface he cites most of the ancient authorities who had argued that the earth was in motion. In an earlier manuscript he even refers to Aristarchus, whose sun-centered universe very closely resembles his own. Although he fails, as was customary during the Renaissance, to mention his more immediate predecessors who had believed that the earth was or could be in motion, he must have known some of their work. He may not, for example, have known of Oresme's contributions, but he had probably at least heard of the very influential treatise in which the fifteenth-century Cardinal, Nicholas of Cusa, derived the motion of the earth from the plurality of worlds in an unbounded Neoplatonic universe. The earth's motion had never been a popular concept, but by the sixteenth century it was scarcely unprecedented. What was unprecedented was the mathematical system that Copernicus built upon the earth's motion. With the possible exception of Aristarchus, Copernicus was the first to realize that the earth's motion might solve an existing astronomical problem or indeed a scientific problem of any sort. Even including Aristarchus, he was the first to develop a detailed account of the astronomical consequences of the earth's motion. Copernicus' mathematics distinguish him from his predecessors, and it was in part because of the mathematics that his work inaugurated a revolution as theirs had not.

Copernicus' Physics and Cosmology

For Copernicus the motion of the earth was a by-product of the problem of the planets. He learned of the earth's motion by examining the celestial motions, and, because the celestial motions had to him a transcendent importance, he was little concerned about the difficulties that his innovation would present to normal men whose concerns were predominantly terrestrial. But Copernicus could not quite ignore the problems that the earth's motion raised for those whose sense of values was less exclusively astronomical than his own.

He had at least to make it possible for his contemporaries to conceive the earth's motion; he had to show that the consequences of this motion were not so devastating as they were commonly supposed to be. Therefore Copernicus opened the *De Revolutionibus* with a non-technical sketch of the universe that he had constructed to house a moving earth. His introductory First Book was directed to laymen, and it included all the arguments that he thought he could make accessible to those without astronomical training.

Those arguments are profoundly unconvincing. Except when they derive from mathematical analyses that Copernicus failed to make explicit in the First Book, they were not new, and they did not quite conform to the details of the astronomical system that Copernicus was to develop in the later books. Only a man who, like Copernicus, had other reasons for supposing that the earth moved could have taken the First Book of the *De Revolutionibus* entirely seriously.

But the First Book is not unimportant. Its very weaknesses foreshadow the incredulity and ridicule with which Copernicus' system would be greeted by those who could not follow the detailed mathematical discussion of the subsequent books. Its repeated dependence upon Aristotelian and scholastic concepts and laws show how little even Copernicus was able to transcend his training and his times except in his own narrow field of specialization. Finally, the incompleteness and incongruities of the First Book illustrate again the coherence of traditional cosmology and traditional astronomy. Copernicus, who was led to revolution by astronomical motives only and who inevitably tried to restrict his innovation to astronomy, could not evade entirely the destructive cosmological consequences of the earth's motion.

BOOK ONE

1. *That the Universe is Spherical.*

In the first place we must observe that the Universe is spherical. This is either because that figure is the most perfect, as not being articulated but whole and complete in itself; or because it is the most capacious and therefore best suited for that which is to contain and preserve all things [of all solids with a given surface the sphere has the greatest volume]; or again because all the perfect parts of it, namely, Sun, Moon and Stars, are so formed; or because all things tend to assume this shape, as is seen in the case of drops of water and liquid bodies in general if freely formed. No one doubts that such a shape has been assigned to the heavenly bodies.

2. *That the Earth also is Spherical.*

The Earth also is spherical, since on all sides it inclines [or falls] toward the center. . . . As we pass from any point northward, the North Pole of the daily rotation gradually rises, while the other pole sinks correspondingly and more stars near the North Pole cease to set, while certain stars in the South do not rise. . . . Further, the change in altitude of the pole is always proportional to the distance traversed on the Earth, which could not be save on a spherical figure. Hence the Earth must be finite and spherical. . . . [Copernicus concludes the chapter with a few more arguments for the earth's sphericity typical of the classical sources that we have already examined.]

3. *How Earth, with the Water on it, forms one Sphere.*

The waters spread around the Earth form the seas and fill the lower declivities. The volume of the waters must be less than that of the Earth, else they would swallow up the land (since both, by their weight, press toward the same center). Thus, for the safety of living things, stretches of the Earth are left uncovered, and also numerous islands widely scattered. Nay, what is a continent, and indeed the whole of the Mainland, but a vast island? . . .

[Copernicus wishes, in this chapter, to show both that the terrestrial globe is predominantly made of earth and that water and earth together are required to make the globe a sphere. Presumably he is looking ahead. Earth breaks up less easily than water when moved; motion of a solid globe is more plausible than of a liquid one. Again, Copernicus will finally say that the earth moves naturally in circles because it is a sphere (see Chapter 8 of his First Book, below). He therefore needs to show that both earth and water are essential to the composition of the sphere, in order that both will participate together in the sphere's natural motion. The passage is of particular interest, because in documenting his view of the structure of the earth Copernicus displays his acquaintance with the recent voyages of discovery and with the corrections that must consequently be made in Ptolemy's geographical writings. For example, he says:

If the terrestrial globe were predominantly water,] the depth of Ocean would constantly increase from the shore outwards, and so neither island nor rock nor anything of the nature of land would be met by sailors, how far soever they ventured. Yet, we know that between the Egyptian Sea and the Arabian Gulf, well-nigh in the middle of the great land-mass, is a passage barely 15 stades wide. On the other hand, in his *Cosmography* Ptolemy would have it that the habitable land extends to the middle circle [of the earth, that is, through a hemisphere extending 180° eastward from the Canary Islands] with a *terra incognita* beyond where modern discovery has added Cathay and a very extensive region as far as 60° of longitude.

Thus we know now that the Earth is inhabited to a greater longitude than is left for Ocean.

This will more evidently appear if we add the islands found in our own time under the Princes of Spain and Portugal, particularly America, a land named after the Captain who discovered it and, on account of its unexplored size, reckoned as another Mainland — besides many other islands hitherto unknown. We thus wonder the less at the so-called Antipodes or Antichthones [the inhabitants of the other hemisphere]. For geometrical argument demands that the Mainland of America on account of its position be diametrically opposite to the Ganges basin in India. . . .

4. *That the Motion of the Heavenly Bodies is Uniform, Circular, and Perpetual, or Composed of Circular Motions.*

We now note that the motion of heavenly bodies is circular. Rotation is natural to a sphere and by that very act is its shape expressed. For here we deal with the simplest kind of body, wherein neither beginning nor end may be discerned nor, if it rotate ever in the same place, may the one be distinguished from the other.

Because there are a multitude of spheres, many motions occur. Most evident to sense is the diurnal rotation . . . marking day and night. By this motion the whole Universe, save Earth alone, is thought to glide from East to West. This is the common measure of all motions, since Time itself is numbered in days. Next we see other revolutions in contest, as it were, with this daily motion and opposing it from West to East. Such opposing motions are those of Sun and Moon and the five planets. . . .

But these bodies exhibit various differences in their motion. First their axes are not that of the diurnal rotation, but of the Zodiac, which is oblique thereto. Secondly, they do not move uniformly even in their own orbits; for are not Sun and Moon found now slower, now swifter in their courses? Further, at times the five planets become stationary at one point and another and even go backward. . . . Furthermore, sometimes they approach Earth, being then in *Perigee*, while at other times receding they are in *Apogee*.

Nevertheless, despite these irregularities, we must conclude that the motions of these bodies are ever circular or compounded of circles. For the irregularities themselves are subject to a definite law and recur at stated times, and this could not happen if the motions were not circular, for a circle alone can thus restore the place of a body as it was. So with the Sun which, by a compounding of circular motions, brings ever again the changing days and nights and the four seasons of the year. Now therein it must be that divers motions are conjoined, since a simple celestial body cannot move irregularly in a single circle. For such irregularity must come of unevenness either in the moving force (whether inherent or acquired) or in the form of the revolving body. Both these alike the mind abhors regarding the most perfectly disposed bodies.

It is then generally agreed that the motions of Sun, Moon, and Planets do but seem irregular either by reason of the divers directions of their axes of revolution, or else by reason that Earth is not the center of the circles in which they revolve, so that to us on Earth the displacements of these bodies [along their orbits] seem greater when they are near [the earth] than when they are more remote (as is demonstrated in optics [or in everyday observation — boats or carriages always seem to move by more quickly when they are closer]). Thus, equal [angular] motions of a sphere, viewed from different distances, will seem to cover different distances in equal times. It is therefore above all needful to observe carefully the relation of the Earth toward the Heavens, lest, searching out the things on high, we should pass by those nearer at hand, and mistakenly ascribe earthly qualities to heavenly bodies.

Copernicus here provides the fullest and most forceful version that we have yet examined of the traditional argument for restricting the motions of celestial bodies to circles. Only a uniform circular motion, or a combination of such motions, can, he thinks, account for the regular recurrence of all celestial phenomena at fixed intervals of time. So far every one of Copernicus' arguments is Aristotelian or scholastic, and his universe is indistinguishable from that of traditional cosmology. In some respects he is even more Aristotelian than many of his predecessors and contemporaries. He will not, for example, consent to the violation of the uniform and symmetric motion of a sphere that is implicit in the use of an equant.

The radical Copernicus has so far shown himself a thoroughgoing conservative. But he cannot postpone the introduction of the earth's motion any longer. He must now take account of his break with tradition. And strangely enough, it is in the break that Copernicus shows his dependence on the tradition most clearly. In dissent he still remains as nearly as possible an Aristotelian. Beginning in the fifth chapter, below, and culminating in the general discussion of motion in the eighth and ninth chapters, Copernicus suggests that because the earth is a sphere, like the celestial bodies, it too must participate in the compounded circular motions which, he says, are natural to a sphere.

5. *Whether Circular Motion belongs to the Earth; and concerning its position.*

Since it has been shown that Earth is spherical, we now consider whether her motion is conformable to her shape and her position in the

Universe. Without these we cannot construct a proper theory of the heavenly phenomena. Now authorities agree that Earth holds firm her place at the center of the Universe, and they regard the contrary as unthinkable, nay as absurd. Yet if we examine more closely it will be seen that this question is not so settled, and needs wider consideration.

A seeming change of place may come of movement either of object or observer, or again of unequal movements of the two (for between equal and parallel motions no movement is perceptible). Now it is Earth from which the rotation of the Heavens is seen. If then some motion of Earth be assumed it will be reproduced in external bodies, which will seem to move in the opposite direction.

Consider first the diurnal rotation. By it the whole Universe, save Earth alone and its contents, appears to move very swiftly. Yet grant that Earth revolves from West to East, and you will find, if you ponder it, that my conclusion is right. It is the vault of Heaven that contains all things, and why should not motion be attributed rather to the contained than to the container, to the located than the locater? The latter view was certainly that of Heraclides and Ecphantus the Pythagorean and Hicetas of Syracuse (according to Cicero). All of them made the Earth rotate in the midst of the Universe, believing that the Stars set owing to the Earth coming in the way, and rise again when it has passed on.

If this [possibility of the earth's motion] is admitted, then a problem no less grave arises about the Earth's position, even though almost everyone has hitherto held that the Earth is at the center of the Universe. [Indeed, if the earth can move at all, it may have more than a simple axial motion about the center of the Universe. It may move away from the center altogether, and there are some good astronomical reasons for supposing that it does.] For grant that Earth is not at the exact center but at a distance from it which, while small compared [with the distance] to the starry sphere, is yet considerable compared with [the distances to] the spheres of the Sun and the other planets. Then calculate the consequent variations in their seeming motions, assuming these [motions] to be really uniform and about some center other than the Earth's. One may then perhaps adduce a reasonable cause for the irregularity of these variable motions. And indeed since the Planets are seen at varying distances from the Earth, the center of Earth is surely not the center of their circles. Nor is it certain whether the Planets move toward and away from Earth, or Earth toward and away from them. It is therefore justifiable to hold that the Earth has another motion in addition to the diurnal rotation. That the Earth, besides rotating, wanders with several motions and is indeed a Planet, is a view attributed to Philolaus the Pythagorean, no mean mathematician, and one whom Plato is said to have sought out in Italy.

Copernicus is here pointing to the most immediate advantage for astronomers of the concept of a moving earth. If the earth moves in an orbital circle around the center as well as spinning on its axis, then, at least qualitatively, the retrograde motions and the different times required for a planet's successive journeys around the ecliptic can be explained without the use of epicycles. In Copernicus' system the major irregularities of the planetary motions are only apparent. Viewed from a moving earth a planet that in fact moved regularly would appear to move irregularly. For this reason, Copernicus feels, we should believe in the orbital motion of the earth. But, strangely enough, in the parts of his work accessible to the lay reader, Copernicus never demonstrates this point any more clearly than he has above. Nor does he demonstrate the other astronomical advantages that he cites elsewhere. He asks the nonmathematical reader to take them for granted, though they are not difficult to demonstrate qualitatively. Only in the later books of the *De Revolutionibus* does he let the real advantages of his system show, and since he there deals, not with retrograde motions in general, but with the abstruse quantitative details of the retrograde motions of each individual planet, only the astronomically initiate were able to discover what the earlier references to astronomical advantages meant. Copernicus' obscurity may have been deliberate, for he had previously referred with some approval to the Pythagorean tradition which dictated withholding nature's secrets from those not previously purified by the study of mathematics (and by other more mystical rites). In any case, the obscurity helps explain the way in which his work was received.

In the next two sections we shall consider the astronomical consequences of the earth's motion in detail, but we must first complete Copernicus' general sketch of physics and cosmology. Omitting for the moment Chapter 6, *Of the Vastness of the Heavens compared with the Size of the Earth*, we proceed to the central chapters in which Copernicus, having asked indulgent readers to assume that astronomical arguments necessitate the earth's motion around the center, attempts to make that motion physically reasonable.

7. Why the Ancients believed that the Earth is at rest, like a Center, in the Middle of the Universe.

The ancient Philosophers tried by divers . . . methods to prove Earth fixed in the midst of the Universe. The most powerful argument was drawn from the doctrine of the heavy and the light. For, they argue, Earth is the heaviest element, and all things of weight move towards it, tending to its center. Hence since the Earth is spherical, and heavy things move vertically to it, they would all rush together to the center if not stopped at the surface. Now those things which move towards the center must, on reaching it, remain at rest. Much more then will the whole Earth remain at rest at the center of the Universe. Receiving all falling bodies, it will remain immovable by its own weight.

Another argument is based on the supposed nature of motion. Aristotle says that motion of a single and simple body is simple. A simple motion may be either straight, or circular. Again a straight motion may be either up or down. So every simple motion must be either toward the center, namely downward, or away from the center, namely upward, or round the center, namely circular. [That is, according to Aristotelian and scholastic physics, natural motions, the only motions that can occur without an external push, are caused by the nature of the body that is in motion. The natural motion of each of the simple bodies (the five elements — earth, water, air, fire, and aether) must itself be simple, because it is a consequence of a simple or elementary nature. And, finally, there are only three (geometrically) simple motions within the spherical universe: up, down, circularly about the center.] Now it is a property only of the heavy elements earth and water to move downward, that is, to seek the center. But the light elements air and fire move upward away from the center. Therefore we must ascribe rectilinear motion to these four elements. The celestial bodies however have circular motion. So far Aristotle.

If then, says Ptolemy, Earth moves at least with a diurnal rotation, the result must be the reverse of that described above. For the motion must be of excessive rapidity, since in 24 hours it must impart a complete rotation to the Earth. Now things rotating very rapidly resist cohesion or, if united, are apt to disperse, unless firmly held together. Ptolemy therefore says that Earth would have been dissipated long ago, and (which is the height of absurdity) would have destroyed the Heavens themselves; and certainly all living creatures and other heavy bodies free to move could not have remained on its surface, but must have been shaken off. Neither could falling objects reach their appointed place vertically beneath, since in the meantime the Earth would have moved swiftly from under them. Moreover clouds and everything in the air would continually move westward. [Note that Copernicus has considerably elaborated Ptolemy's original argument, quoted on p. 85. It is by no means clear that Ptolemy would have gone this far.]

8. *The Insufficiency of these Arguments, and their Refutation.*

For these and like reasons, they say that Earth surely rests at the center of the Universe. Now if one should say that the Earth *moves*, that is as much as to say that the motion is natural, not violent [or due to an external push]; and things which happen according to nature produce the opposite effects to those which occur by violence. Things subjected to any force or impetus, gradual or sudden, must be disintegrated, and cannot long exist. But natural processes being adapted to their purpose work smoothly. [That is, if the earth moves at all, it does so because it is of the nature of earth to move, and a natural motion cannot be disruptive.]

Idle therefore is the fear of Ptolemy that Earth and all thereon would be disintegrated by a natural rotation, a thing far different from an artificial act. Should he not fear even more for the Universe, whose motion must be as much more rapid as the Heavens are greater than the Earth? Have the Heavens become so vast because of their vehement motion, and would they collapse if they stood still? If this were so the Heavens must be of infinite size. For the more they expand by the force of their motion, the more rapid will become the motion because of the ever increasing distance to be traversed in 24 hours. And in turn, as the motion waxes, must the immensity of the Heavens wax. Thus velocity and size would increase each the other to infinity. . . .

They say too that outside the Heavens is no body, no space, nay not even void, in fact absolutely nothing, and therefore no room for the Heavens to expand [as we have suggested above that they would]. Yet surely it is strange that something can be held by nothing. Perhaps indeed it will be easier to understand this nothingness outside the Heavens if we assume them to be infinite, and bounded internally only by their concavity, so that everything, however great, is contained in them, while the Heavens remain immovable. . . .

Let us then leave to Natural Philosophers the question whether the Universe be finite or no, holding only to this that Earth is finite and spherical. Why then hesitate to grant Earth that power of motion natural to its [spherical] shape, rather than suppose a gliding round of the whole universe, whose limits are unknown and unknowable? And why not grant that the diurnal rotation is only apparent in the Heavens but real in the Earth? It is but as the saying of Aeneas in Virgil — "We sail forth from the harbor, and lands and cities retire." As the ship floats along in the calm, all external things seem to have the motion that is really that of the ship, while those within the ship feel that they and all its contents are at rest.

It may be asked what of the clouds and other objects suspended in the air, or sinking and rising in it? Surely not only the Earth, with the water on it, moves thus, but also a quantity of air and all things so associated with the Earth. Perhaps the contiguous air contains an admixture of earthy or watery matter and so follows the same natural law as the Earth, or perhaps the air acquires motion from the perpetually rotating Earth by propinquity and absence of resistance. . . .

We must admit the possibility of a double motion of objects which fall and rise in the Universe, namely the resultant of rectilinear and circular motion. [This is the analysis advocated earlier by Oresme.] Thus heavy falling objects, being specially earthy, must doubtless retain the nature of the whole to which they belong. . . . [Therefore a stone, for example, when removed from the earth will continue to move circularly with the earth and will simultaneously fall rectilinearly toward the earth's surface. Its net motion will be some sort of spiral, like the motion of a bug that crawls straight toward the center of a rotating potter's wheel.]

That the motion of a simple body must be simple is true then primarily of circular motion, and only so long as the simple body rests in its own natural place and state. In that state no motion save circular is possible, for such motion is wholly self-contained and similar to being at rest. But if objects move or are moved from their natural place rectilinear motion supervenes. Now it is inconsistent with the whole order and form of the Universe that it should be outside its own place. Therefore there is no rectilinear motion save of objects out of their right place, nor is such motion natural to perfect objects, since [by such a motion] they would be separated from the whole to which they belong and thus would destroy its unity. . . . [Copernicus' argument shows how quickly the traditional distinction between the terrestrial and the celestial regions must disappear when the earth becomes a planet, for he is here simply applying a traditional argument about celestial bodies to the earth. Circular motion, whether simple or compound, is the nearest thing to rest. It can be natural to the earth just as it has always been natural to the heavens, because it cannot disrupt the observed unity and regularity of the universe. Linear motion, on the other hand, cannot be natural to any object that has achieved its own place, for linear motion is disruptive and a natural motion that destroys the universe is absurd.]

Further, we conceive immobility to be nobler and more divine than change and inconstancy, which latter is thus more appropriate to Earth than to the Universe. Would it not then seem absurd to ascribe motion to that which contains or locates, and not rather to that contained and located, namely the Earth?

Lastly, since the planets approach and recede from the Earth, both their motion round the center, which is held [by Aristotelians] to be the Earth, and also their motion outward and inward are the motion of one body. [And this violates the very laws from which Aristotelians derive the central position of the earth, for according to these laws the planets should have only a single motion.] Therefore we must accept this motion round the center in a more general sense, and must be satisfied provided that every motion has a proper center. From all these considerations it is more probable that the Earth moves than that it remains at rest. This is especially the case with the diurnal rotation, as being particularly a property of the Earth.

9. *Whether more than one Motion can be attributed to the Earth, and of the center of the Universe.*

Since then there is no reason why the Earth should not possess the power of motion, we must consider whether in fact it has more motions than one, so as to be reckoned as a Planet.

That Earth is not the center of all revolutions is proved by the apparently irregular motions of the planets and the variations in their distances from the Earth. These would be unintelligible if they moved in circles concentric with Earth. Since, therefore, there are more centers than one [that is, a center for all the orbital motions, a center of the earth itself, and perhaps others besides], we may discuss whether the center of the Universe is or is not the Earth's center of gravity.

Now it seems to me gravity is but a natural inclination, bestowed on the parts of bodies by the Creator so as to combine the parts in the form of a sphere and thus contribute to their unity and integrity. And we may believe this property present even in the Sun, Moon, and Planets, so that thereby they retain their spherical form notwithstanding their various paths. If, therefore, the Earth also has other motions, these must necessarily resemble the many outside [planetary] motions having a yearly period [since the earth now seems like a planet in so many other respects]. For if we transfer the motion of the Sun to the Earth, taking the Sun to be at rest, then morning and evening risings and settings of Stars will be unaffected, while the stationary points, retrogressions, and progressions of the Planets are due not to their own motions, but to that of the Earth, which their appearances reflect. Finally we shall place the Sun himself at the center of the Universe. All this is suggested by the systematic procession of events and the harmony of the whole Universe, if only we face the facts, as they say, "with both eyes open."

In these last three chapters we have Copernicus' theory of motion, a conceptual scheme that he designed to permit his transposing the earth and sun without tearing apart an essentially Aristotelian universe in the process. According to Copernicus' physics all matter, celestial and terrestrial, aggregates naturally into spheres, and the spheres then rotate of their own nature. A bit of matter separated from its natural position will continue to rotate with its sphere, simultaneously returning to its natural place by a rectilinear motion. It is a singularly incongruous theory (as Chapter 6 will demonstrate in more detail), and, in all but its most incongruous portions, it is a relatively unoriginal one. Copernicus may possibly have reinvented it for himself, but most of the essential elements in both his criticism of Aristotle and his theory of motion can be found in earlier scholastic writers, particularly in Oresme. Only when applied to Oresme's more limited problem, they are less implausible.

Failure to provide an adequate physical basis for the earth's motion does not discredit Copernicus. He did not conceive or accept the earth's motion for reasons drawn from physics. The physical and cosmological problem treated so crudely in the First Book are of his making, but they are not really his problems; he might have avoided them altogether if he could. But the inadequacies of Copernicus' physics do illustrate the way in which the consequences of his astronomical innovation transcend the astronomical problem from which the innovation was derived, and they do show how little the author of the innovation was himself able to assimilate the Revolution born from his work. The moving earth is an anomaly in a classical Aristotelian universe, but the universe of the *De Revolutionibus* is classical in every respect that Copernicus can make seem compatible with the motion of the earth. As he says himself, the motion of the sun has simply been transferred to the earth. The sun is not yet a star but the unique central body about which the universe is constructed; it inherits the old functions of the earth and some new ones besides. As we shall soon discover, Copernicus' universe is still finite, and concentric nesting spheres still move all planets, even though they can no longer be driven by the outer sphere, which is now at rest. All motions must be compounded of circles; moving the earth does not even enable Copernicus to dispense with epicycles. The Copernican Revolution, as we know it, is scarcely to be found in the *De Revolutionibus* and that is the second essential incongruity of the text.

Copernican Astronomy — The Two Spheres

We have not quite finished with Copernicus' First Book. But Chapters 10 and 11, which immediately follow the last section quoted above, deal with more nearly astronomical matters, and we shall consider them in the context of an astronomical discussion which goes beyond the arguments that Copernicus made accessible for lay readers. We shall again turn briefly to Copernicus' text in a later section, but first we shall try to discover why astronomers might have been more impressed than laymen with Copernicus' proposal. That can scarcely be discovered anywhere in the First Book.

Copernicus endowed the earth with three simultaneous circular motions: a diurnal axial rotation, an annual orbital motion, and an annual conical motion of the axis. The eastward diurnal rotation is the one that accounts for the apparent diurnal circles traced by the stars, sun, moon, and planets. If the earth is situated at the center of the sphere of the stars, and rotates eastward daily about an axis through its own north and south poles, then all objects that are stationary or nearly stationary with respect to the sphere of the stars will seem to travel westward in circular arcs above the horizon, arcs just like those in which the celestial bodies are observed to move in any short period of time.

If Copernicus' or Oresme's arguments to this effect are obscure, refer again to the star trails shown in Figures 6 and 7 (pp. 18 and 19). Those tracks could be produced either by a circular motion of the stars in front of a fixed observer (Ptolemy's explanation) or by a rotation of the observer in front of fixed stars (Copernicus' explanation). Or examine the new two-sphere universe shown in Figure 26, a

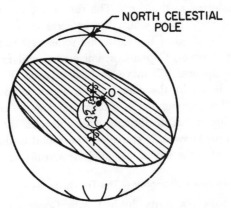

Figure 26. A rotating earth at the center of a fixed stellar sphere. In comparing this diagram with Figure 11, notice that here the horizon plane must be turned with the earth, so that its geometric relation to the moving observer O stays fixed.

simplified copy of the drawing that we first used in discussing the motions of stars in the two-sphere universe (Figure 11, p. 31) except that in the new version the poles are shown for the earth, not for the celestial sphere, and the direction of rotation has been reversed. When we first used a diagram like this, we held the earth, the observer, and the horizon plane fixed, and we turned the sphere of the stars westward. Now we must hold the outer sphere fixed and spin the earth, observer, and horizon plane together eastward. An observer sitting at the center of the horizon plane and moving with it will not be able to tell, at least from anything he can see in the skies, any difference between the two cases. In both he will see stars and planets emerge along the eastern rim of the horizon and travel overhead to the western horizon in the same circular paths.

To this point we have kept the spinning earth at the center of the stationary sphere of the stars; we have, that is, considered the model of the universe suggested by Heraclides and developed by Oresme. This is only the first step toward a Copernican universe, however, and the next one is both more radical and more difficult. As Copernicus points out in the portion of Chapter 5 already quoted, if we are prepared to admit the possibility of the earth's motion at all, we must be prepared to consider not only a motion at the center, but also a motion of the earth away from the center. In fact, says Copernicus, a moving earth need not be at the center. It need only be relatively near the center, and as long as it stays close enough to the center it may move about at will without affecting the apparent motions of the stars. This was a difficult conclusion for his astronomically trained colleagues to accept because, in contrast to the conception of the earth's immobility, which derives only from common sense and from terrestrial physics, the notion of the earth's central position can apparently be derived directly from astronomical observation.) Copernicus' conception of a noncentral earth therefore seemed initially to conflict with the immediate consequences of pure astronomical observation, and it was to avoid this conflict, or a closely related one which we shall consider at the end of the next section, that Copernicus was forced to increase vastly the size of the sphere of the stars and to take a first step toward the conception of an infinite universe elaborated by his successors. Copernicus' discussion of the earth's position occurs in Chapter 6 of his First Book. Here we shall need a clearer and more comprehensive version.

The earth's central position within the sphere of the stars can apparently be derived from the observation that the horizon of any terrestrial observer bisects the stellar sphere. The vernal equinox and the autumnal equinox are, for example, two diametrically opposite points on the sphere of the stars, for they are defined as the intersections of two great circles on the sphere, the equator and the ecliptic. Observation shows that whenever one of these points is just rising over the horizon on the east, the other is just setting in the west. The same is true of any other pair of diametrically opposite points on the sphere: whenever one rises, the other sets. Apparently these observations can be explained only if, as shown in Figure 26 or the earlier Figure 11, the horizon plane is drawn through the center of the sphere of the stars so that it, too, will intersect the sphere in a great circle. If and only if the horizon plane intersects the sphere of the stars in a great circle will diametrically opposite points on the sphere always rise and set at the same moment.

But all horizon planes must also be drawn tangent to the spherical earth. (We have avoided this construction in Figures 26 and 11 only because we have there shown the earth immensely exaggerated in size.) Therefore the observer must himself be at, or very nearly at, the center of the sphere of the stars. The entire surface of the terrestrial sphere itself must be at or very nearly at the center; the earth must be very small, almost a point, and it must be centrally located. If, as in Figure 27, the earth (represented by the inner concentric

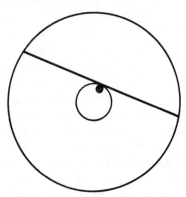

Figure 27. If the earth's diameter is appreciable compared with that of the sphere of the stars or if the earth is appreciably displaced from the center, the horizon plane does not bisect the stellar sphere.

circle) were quite large with respect to the sphere of the stars or if the earth (now represented by the black dot) were small but displaced from the center, then the horizon plane would not seem to bisect the sphere of the stars, and diametrically opposite points on the sphere would not rise and set together.

As developed here the argument itself makes clear the weakness exploited by Copernicus. Observation does not show that the earth must be a point (if it did, even the Aristotelian and Ptolemaic universe would conflict with observation) or that it must be precisely at the center, because observation can never say that, for example, the vernal equinox rises *exactly* as the autumnal equinox sets. Crude naked-eye observations will show that when the vernal equinox is just setting, the autumnal equinox is within a degree or so of the horizon. Refined naked-eye observation (appropriately corrected for atmospheric refraction and for the irregularities of any actual horizon) might show that when the winter solstice has just reached the western horizon, the summer solstice is within 6' (or 0.1°) of the eastern horizon. But no naked-eye observation will do much better. It can show only that the horizon *very nearly* bisects the sphere and that all terrestrial observers must therefore be very close to the center of the universe. Just how nearly the horizon bisects the sphere and just how close to the center terrestrial observers must be depends upon the accuracy of observation.

For example, if we know from observation that whenever one solstice lies on the horizon, the other is *no more than* 0.1° away from the horizon, then no terrestrial observer may ever be farther from the center of the sphere of the stars than a distance which is 0.001 the radius of that sphere. Or if observation tells us (and few naked-eye observations are even approximately this good) that with one solstice on the horizon the other is no more than 0.01° away from the horizon, then the inner sphere of Figure 27 may have a radius no larger than 0.0001 the radius of the outer sphere, and the entire earth must again lie somewhere within the inner circle at all times. If the earth moved outside the inner circle, then the horizon plane would fail to bisect the sphere of the stars by more than 0.01°, and our hypothetical observations would discover the discrepancy, but with the earth anywhere inside of the inner circle, the horizon plane will seem, within the limits of observation, to bisect the sphere.

That is Copernicus' argument. Observation only forces us to keep the earth somewhere inside of a small sphere concentric with the sphere of the stars. Within that inner sphere the earth may move freely without violating the appearances. In particular, the earth may have an orbital motion about the center or about the central sun, provided that its orbit never carries it too far away from the center. And "too far" means only "too far relative to the radius of the outer sphere." If the radius of the outer sphere is known, then observations of known accuracy place a limit upon the *maximum* radius of the earth's orbit. If the size of the earth's orbit is known (and it can, in theory, be determined by Aristarchus' technique for measuring the earth-sun distance), then observations of known accuracy place a limit upon the *minimum* size of the sphere of the stars. For example, if the distance between the earth and sun is, as indicated by Aristarchus' measurement described in the Technical Appendix, equal to 764 earth diameters (1528 earth radii) and if observations are known to be accurate within 0.1°, then the radius of the sphere of the stars must be at least 1000 times the radius of the earth's orbit or at least 1,528,000 earth radii.

Our example is a useful one, because, though Copernicus' observations were not quite this accurate, those made by his immediate successor, Brahe, were if anything slightly more accurate than 0.1°. Ours is a representative estimate of the minimum size of the sphere of the stars by a sixteenth-century Copernican. In principle, there is nothing absurd about the result, for in the sixteenth and seventeenth centuries there was no direct way of determining the distance to the sphere of the stars. Its radius might have been more than 1,500,000 earth radii. But if it were that large — and Copernicanism demanded that it should be — then a real break with traditional cosmology must be admitted. Al Fargani, for example, had estimated the radius of the sphere as 20,110 earth radii, more than seventy-five times smaller than the Copernican estimate. The Copernican universe must be vastly larger than that of traditional cosmology. Its volume is *at least* 400,000 times as great. There is an immense amount of space between the sphere of Saturn and the sphere of the stars. The neat functional coherence of the nesting spheres of the traditional universe has been violated, though Copernicus seems to remain sublimely unaware of the break.

Copernican Astronomy — The Sun

Copernicus' argument permits an orbital motion of the earth in a vastly expanded universe, but the point is academic unless the orbital motion can be shown to be compatible with the observed motions of the sun and other planets. It is to those motions that Copernicus turns in Chapters 10 and 11 of his First Book. We may best begin with an expanded paraphrase of Chapter 11, in which

Copernicus describes the orbital motion of the earth and considers its effect upon the apparent position of the sun. For the moment assume, as shown in Figure 28, that the centers of the universe, the sun, and the earth's orbit all coincide. In the diagram the plane of the ecliptic is viewed from a position near the north celestial pole; the sphere of the stars is stationary; the earth travels regularly eastward in its orbit once in a year; and it simultaneously spins eastward on its axis once in every 23 hours 56 minutes. Provided that the earth's orbit is much

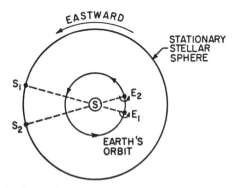

Figure 28. As the earth moves in its Copernican orbit from E_1 to E_2, the apparent position of the central sun, S, seen against the sphere of the stars shifts from S_1 to S_2.

smaller than the sphere of the stars, the axial rotation of the earth will account precisely for the diurnal circles of the sun, moon, and planets, as well as for those of the stars, because from any position in the earth's orbit all of these bodies must be seen against the sphere of the stars and must seem to move with it as the earth rotates.

In the diagram the earth is shown in two positions which it occupies thirty days apart. In each position the sun is viewed against the sphere of the stars, and both apparent positions of the sun must lie on the ecliptic, which is now defined as the line in which the plane of the earth's motion (a plane that includes the sun) intersects the sphere. But as the earth has moved eastward from position E_1 to position E_2 in the diagram, the sun has apparently moved eastward along the ecliptic from position S_1 to position S_2. Copernicus' theory therefore predicts just the same eastward annual motion of the sun along the ecliptic as the Ptolemaic theory. It also predicts, as we shall discover immediately, the same seasonal variation of the height of the sun in the sky.

Figure 29 shows the earth's orbit viewed from a point in the celestial sphere slightly north of the autumnal equinox. The earth is drawn at the four positions occupied successively at the vernal equinox, the summer solstice, the autumnal equinox, and the winter solstice. In all four of these positions, as throughout its motion, the earth's axis remains parallel to an imaginary line passing through the sun and tilted 23½° from a perpendicular to the plane of the ecliptic. Two

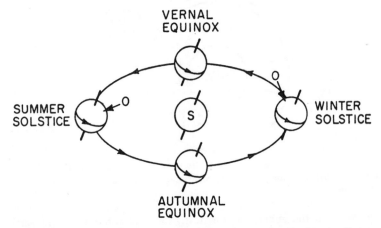

Figure 29. The earth's annual motion around its Copernican orbit. At all times the earth's axis stays parallel to itself or to the stationary line drawn through the sun. As a result an observer *O* at noon in middle-northern latitudes finds the sun much more nearly overhead at the summer than at the winter solstice.

little arrows in the diagram show the position of a terrestrial observer in middle-northern latitudes at local noon on June 22 and December 22, the two solstices. Lines from the sun to the earth (not shown in the diagram) indicate the direction of the rays of the noon sun, which is clearly more nearly over the observer's head during the summer solstice than during the winter solstice. A similar construction determines the sun's elevation at the equinoxes and at intermediate seasons.

The seasonal variation of the sun's elevation can therefore be completely diagnosed from Figure 29. In practice, however, it is simpler to revert to the Ptolemaic explanation. Since in every season the sun appears to occupy the same position among the stars in the Copernican as in the Ptolemaic system, it must rise and set with the same stars in both systems. The correlation of the seasons with the apparent position of the sun along the ecliptic cannot be affected by the transition. With respect to the apparent motions of the sun and stars the two systems are equivalent, and the Ptolemaic is simpler.

The last diagram also reveals two other interesting features of Copernicus' system. Since it is the rotation of the earth that produces the diurnal circles of the stars, the earth's axis must point to the center of those circles in the celestial sphere. But, as the diagram indicates, the earth's axis never does point to quite the same positions on the celestial sphere from one year's end to the next. According to the Copernican theory the extension of the earth's axis traces, during the course of a year, two small circles on the sphere of the stars, one around the north celestial pole and one around the south. To an observer on the earth the center of the diurnal circles of the stars should itself seem to move in a small circle about the celestial pole once each year. Or, to put the same point in a way more closely related to observation, each of the stars should seem slightly to change its position on the sphere of the stars (or with respect to the observed pole of the sphere) during the course of a year.

This apparent motion, which cannot be seen with the naked eye and which was not even seen with telescopes until 1838, is known as

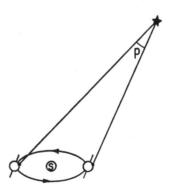

Figure 30. The annual parallax of a star. Because the line between a terrestrial observer and a fixed star does not stay quite parallel to itself as the earth moves in its orbit, the star's apparent position on the stellar sphere should shift by an angle p during an interval of six months.

the parallactic motion. Because two lines drawn to a star from diametrically opposite points on the earth's orbit are not quite parallel
(Figure 30), the apparent angular position of the star viewed from
the earth should be different at different seasons. But if the distance to
the star is very much greater than the distance across the earth's
orbit, then the angle of parallax, p in Figure 30, will be very, very
small, and the change in the apparent position of the star will not
be appreciable. The parallactic motion is not apparent only because
the stars are so very far away relative to the dimensions of the earth's
orbit. The situation is precisely equivalent to the one we discussed
above when considering why the earth's motion did not seem to
change the intersection of the horizon plane and the sphere of the
stars. In fact, we are dealing with the same problem. But the present
version of the problem is a more important one, because near the
horizon it is very difficult to make the precise measurements of stellar
position required to discover whether the horizon bisects the stellar
sphere. Unlike the rising and setting of the equinoxes, discussed above,
the search for parallactic motions need not be restricted to the horizon.
Parallax therefore provides a much more sensitive observational check
upon the minimum size of the sphere of the stars relative to the size
of the earth's orbit than is provided by the position of the horizon,
and the Copernican estimates of the sphere's size given above ought
really to have been derived from a discussion of parallax.

The second point illuminated by considering Figure 29 is not
about the skies at all but about Copernicus. We described the orbital
motion illustrated in the diagram as a single motion by which the
earth's center is carried in a circle about the sun while its axis remains always parallel to a fixed line through the sun. Copernicus describes the same physical motion as consisting of two simultaneous
mathematical motions. That is why he gives the earth a total of three
circular motions. And the reasons for his description give another significant illustration of the extent to which his thought was bound to
the traditional patterns of Aristotelian thought. For him the earth
is a planet which is carried about the central sun by a sphere just
like the one that used to carry the sun about the central earth. If the
earth were firmly fixed in a sphere, its axis would not always stay
parallel to the same line through the sun; it would instead be carried
about by the sphere's rotation and would occupy the positions shown

in Figure 31a. After the earth had revolved 180° about the sun, the earth's axis would still be tilted 23½° away from the perpendicular but in a direction opposite to that in which it had begun. To undo this change in the direction of the axis, caused by the rotation of the

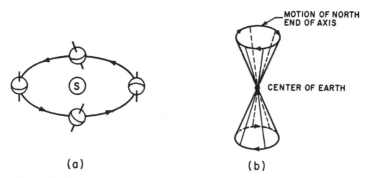

Figure 31. Copernicus' "second" and "third" motions. The second motion, that of a planet fixed in a rotating sun-centered sphere, is shown in (a). This motion does not keep the earth's axis parallel to itself, so that the conical third motion shown in (b) is required to bring the axis back into line.

sphere that carries the earth, Copernicus requires a third circular motion, this one applied to the axis of the earth only and shown in Figure 31b. It is a conical motion, which carries the north end of the axis once westward each year, and thus just compensates for the effect on the earth's axis of the orbital motion.

Copernican Astronomy — The Planets

So far the conceptual scheme developed by Copernicus is just as effective as Ptolemy's, but it is surely no more so, and it seems a good deal more cumbersome. It is only when the planets are added to Copernicus' universe that any real basis for his innovation becomes apparent. Consider, for example, the explanation of retrograde motion to which Copernicus alluded without discussion at the end of Chapter 5 in his introductory First Book. In the Ptolemaic system the retrograde motion of each planet is accounted for by placing the planet on a major epicycle whose center is, in turn, carried about the earth by the planet's deferent. The combined motion of these two circles produces the characteristic looped patterns discussed in Chapter 3.

In Copernicus' system no major epicycles are required. The retrograde or westward motion of a planet among the stars is only an apparent motion, produced, like the apparent motion of the sun around the ecliptic, by the orbital motion of the earth. According to Copernicus the motion that Ptolemy had explained with major epicycles was really the motion of the earth, attributed to the planets by a terrestrial observer who thought himself stationary.

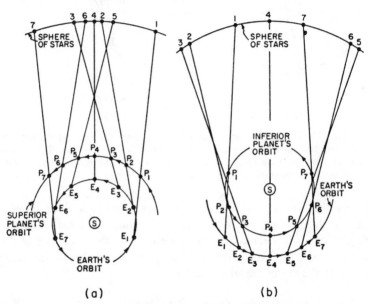

Figure 32. The Copernican explanation of retrograde motion for (a) superior planets and (b) inferior planets. In each diagram the earth moves steadily on its orbit from E_1 to E_7 and the planet moves from P_1 to P_7. Simultaneously the planet's apparent position against the stellar sphere shifts eastward from 1 to 7, but as the two planets pass there is a brief westward retrogression from 3 to 5.

The basis of Copernicus' contention is illustrated and clarified by Figures 32a and 32b. Successive apparent positions of a moving superior planet viewed from a moving earth against the fixed background provided by the stellar sphere are shown in the first diagram; the second shows successive apparent positions of an inferior planet. Only the orbital motions are indicated; the earth's diurnal rotation, which produces the rapid apparent westward motion of the sun, planets, and stars together, is omitted. In both diagrams successive positions of the earth in its sun-centered circular orbit are indicated by the points $E_1, E_2, \ldots, E_7$; the corresponding consecutive positions of the planets are marked $P_1, P_2, \ldots, P_7$; and the corresponding apparent positions of the planet, discovered by extending a line from the earth through the planet until it intersects the stellar sphere, are labeled 1, 2, . . . , 7. In each case the more central planet moves more rapidly in its orbit. Inspection of the diagram indicates that the apparent motion of the planet among the stars is normal (eastward) from 1 to 2 and from 2 to 3; then the planet appears to retrogress (move westward) from 3 to 4 and from 4 to 5; and finally it reverses its motion again and moves normally from 5 to 6 and from 6 to 7. As the earth completes the balance of its orbit, the planet continues in normal motion, moving eastward most rapidly when it lies diametrically across the sun from the earth.

Therefore, in Copernicus' system, planets viewed from the earth should appear to move eastward most of the time; they retrogress only when the earth, in its more rapid orbital motion, overtakes them (superior planets) or when they overtake the earth (inferior planets). Retrograde motion can occur only when the earth is nearest to the planet whose motion is observed, and this is in accord with observations. Superior planets, at least, are most brilliant when they move westward. The first major irregularity of planetary motion has been explained qualitatively without the use of epicycles.

Figure 33 indicates how Copernicus' proposal accounts for a second major irregularity of the planetary motions — the discrepancy between the times required for successive trips of a planet around the ecliptic. In the diagram it is assumed that the earth completes $1\frac{1}{4}$ eastward trips about its orbit while the planet, in this case a superior planet, travels eastward through its orbit once. Suppose that at the start of the series of observations the earth is at E_1 and the planet at P. The planet is then in the middle of a retrogression and appears silhouetted against the stationary stellar sphere at 1. When the planet has completed one revolution in its orbit and returned to P, the earth has made $1\frac{1}{4}$ trips around its orbit and reached E_2. The planet therefore is seen at 2, west of position 1 at which it started. It has not yet completed a full journey around the ecliptic, and its first full trip will therefore consume more time than the planet required to revolve once in its orbit.

As the planet makes its second trip about its orbit, the earth again makes more than one orbital revolution and reaches E_3 when the planet has returned to P again. This time the planet is seen silhouetted at 3, to the east of position 2. It has completed more than one journey around the ecliptic while moving only once through its orbit, and its second journey around the ecliptic was therefore a very rapid one. After a third revolution the planet is again at P, but it appears at position 4, east of 3, and its journey around the ecliptic was therefore

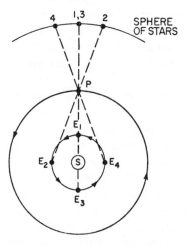

Figure 33. The Copernican explanation of variations in the time required for a superior planet to complete successive journeys around the ecliptic. While the planet moves once eastward around its orbit from P to P, the earth makes 1¼ eastward revolutions from E_1 to E_1 and on to E_2. During this interval the apparent position of the planet among the stars moves eastward from 1 to 2, slightly less than a full trip. During the planet's next revolution the earth moves from E_2 to E_2 and on to E_3, so that its apparent position among the stars shifts from 2 to 1 and on to 1 again, slightly more than one full trip around the ecliptic.

again a fast one. After a fourth revolution in its orbit the planet again appears at 1, west of 4, and its final trip was therefore slow. The planet has completed four trips about its orbit and four trips around the ecliptic at the same instant. The average time required by a superior planet to circle the ecliptic is therefore identical with the planet's orbital period. But the time required for an individual trip may be considerably greater or considerably less than the average. A similar argument will account for the similar irregularities of an inferior planet's motion.

Retrograde motion and the variation of the time required to circle the ecliptic are the two gross planetary irregularities which in antiquity had led astronomers to employ epicycles and deferents in treating the problem of the planets. Copernicus' system explains these same gross irregularities, and it does so without resorting to epicycles, or at least to major epicycles. To gain even an approximate and qualitative account of the planetary motions Hipparchus and Ptolemy had required twelve circles — one each for the sun and moon, and two each for the five remaining "wanderers." Copernicus achieved the same qualitative account of the apparent planetary motions with only seven circles. He needed only one sun-centered circle for each of the six known planets — Mercury, Venus, Earth, Mars, Jupiter, and Saturn — and one additional earth-centered circle for the moon. To an astronomer concerned only with a qualitative account of the planetary motions, Copernicus' system must seem the more economical.

But this apparent economy of the Copernican system, though it is a propaganda victory that the proponents of the new astronomy rarely failed to emphasize, is largely an illusion. We have not yet begun to deal with the full complexity of Copernicus' planetary astronomy. The seven-circle system presented in the First Book of the *De Revolutionibus*, and in many modern elementary accounts of the Copernican system, is a wonderfully economical system, but it does not work. It will not predict the position of planets with an accuracy comparable to that supplied by Ptolemy's system. Its accuracy is comparable to that of a simplified twelve-circle version of Ptolemy's system — Copernicus can give a more economical *qualitative* account of the planetary motions than Ptolemy. But to gain a reasonably good *quantitative* account of the alteration of planetary position Ptolemy had been compelled to complicate the fundamental twelve-circle system with minor epicycles, eccentrics, and equants, and to get comparable results from his basic seven-circle system Copernicus, too, was forced to use minor epicycles and eccentrics. His full system was little if any less cumbersome than Ptolemy's had been. Both employed over thirty circles; there was little to choose between them in economy. Nor could the two systems be distinguished by their accuracy. When Copernicus had finished adding circles, his cumbersome sun-centered system gave results as accurate as Ptolemy's, but it did not give more accurate results. Copernicus did not solve the problem of the planets.

The full Copernican system is described in the latter books of the *De Revolutionibus*. Fortunately we need only illustrate the sorts of complexities there developed. Copernicus' system was not, for example, really a sun-centered system at all. To account for the increased rate at which the sun travels through the signs of the zodiac during the winter, Copernicus made the earth's circular orbit eccentric, displacing its center from the sun's. To account for other irregularities, indicated by ancient and contemporary observations of the sun's motion, he kept this displaced center in motion. The center of the earth's eccentric was placed upon a second circle whose motion continually varied the extent and direction of the earth's eccentricity. The final system employed to compute the earth's motion is represented approximately in Figure 34a. In the diagram, S is the sun, fixed in space; the point O, which itself moves slowly about the sun, is the center of a slowly rotating circle that carries the moving center O_E of the earth's eccentric; E is the earth itself.

Similar complexities were necessitated by the observed motions of the other heavenly bodies. For the moon Copernicus used a total of three circles, the first centered on the moving earth, the second centered on the moving circumference of the first, and the third on the

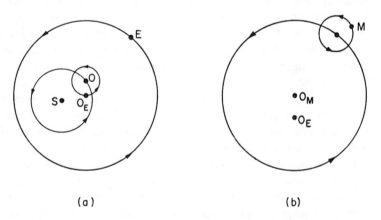

(a)   (b)

Figure 34. Copernicus' account of the motion of (*a*) the earth and (*b*) Mars. In (*a*) the sun is at S, and the earth, E, revolves on a circle whose center, O_E, revolves slowly about a point O, which in turn revolves on a sun-centered circle. In (*b*) Mars is placed on an epicycle revolving on a deferent whose center, O_M, maintains a fixed geometric relation to the moving center O_E of the earth's orbit.

circumference of the second. For Mars and most of the other planets he employed a system much like that illustrated in Figure 34b. The center of Mars's orbit, O_M, is displaced from the center of the earth's orbit, O_E, and is moved with it; the planet itself is placed at M, not on the eccentric but on an epicycle, which rotates eastward in the same direction and with the same period as the eccentric. Nor do the complexities end here. Still other devices, fully equivalent to Ptolemy's, were required to account for the north and south deviations of each planet from the ecliptic.

Even this brief sketch of the complex system of interlocking circles employed by Copernicus to compute planetary position indicates the third great incongruity of the *De Revolutionibus* and the immense irony of Copernicus' lifework. The preface to the *De Revolutionibus* opens with a forceful indictment of Ptolemaic astronomy for its inaccuracy, complexity, and inconsistency, yet before Copernicus' text closes, it has convicted itself of exactly the same shortcomings. Copernicus' system is neither simpler nor more accurate than Ptolemy's. And the methods that Copernicus employed in constructing it seem just as little likely as the methods of Ptolemy to produce a single consistent solution of the problem of the planets. The *De Revolutionibus* itself is not consistent with the single surviving early version of the system, described by Copernicus in the early manuscript *Commentariolus*. Even Copernicus could not derive from his hypothesis a single and unique combination of interlocking circles, and his successors did not do so. Those features of the ancient tradition which had led Copernicus to attempt a radical innovation were not eliminated by that innovation. Copernicus had rejected the Ptolemaic tradition because of his discovery that "the Mathematicians are inconsistent in these [astronomical] investigations" and because "if their hypotheses were not misleading, all inferences based thereon might surely be verified." A new Copernicus could have turned the identical arguments against him.

The Harmony of the Copernican System

Judged on purely practical grounds, Copernicus' new planetary system was a failure; it was neither more accurate nor significantly simpler than its Ptolemaic predecessors. But historically the new system was a great success; the *De Revolutionibus* did convince a few of Copernicus' successors that sun-centered astronomy held the key to the problem of the planets, and these men finally provided the simple and accurate solution that Copernicus had sought. We shall examine their work in the next chapter, but first we must try to discover why they became Copernicans — in the absence of increased economy or precision, what reasons were there for transposing the earth and the sun? The answer to this question is not easily disentangled from the technical details that fill the *De Revolutionibus*, because, as Copernicus himself recognized, the real appeal of sun-centered astronomy was aesthetic rather than pragmatic. To astronomers the initial choice between Copernicus' system and Ptolemy's could only be a matter of taste, and matters of taste are the most difficult of all to define or debate. Yet, as the Copernican Revolution itself indicates, matters of taste are not negligible. The ear equipped to discern geometric harmony could detect a new neatness and coherence in the sun-centered astronomy of Copernicus, and if that neatness and coherence had not been recognized, there might have been no Revolution.

We have already examined one of the aesthetic advantages of Copernicus' system. It explains the principal *qualitative* features of the planetary motions without using epicycles. Retrograde motion, in particular, is transformed to a natural and immediate consequence of the geometry of sun-centered orbits. But only astronomers who valued qualitative neatness far more than quantitative accuracy (and there were a few — Galileo among them) could consider this a convincing argument in the face of the complex system of epicycles and eccentrics elaborated in the *De Revolutionibus*. Fortunately there were other, less ephemeral, arguments for the new system. For example, it gives a simpler and far more natural account than Ptolemy's of the motions of the inferior planets. Mercury and Venus never get very far from the sun, and Ptolemaic astronomy accounts for this observation by tying the deferents of Mercury, Venus, and the sun together so that the center of the epicycle of each inferior planet always lies on a straight line between the earth and the sun (Figure 35*a*). This alignment of the centers of the epicycles is an "extra" device, an *ad hoc* addition to the geometry of earth-centered astronomy, and there is no need for such an assumption in Copernicus' system. When, as in

Figure 35*b*, the orbit of a planet lies entirely within the earth's orbit, there is no way in which the planet can appear far from the sun. Maximum elongation will occur when, as in the diagram, the line from the earth to the planet is tangent to the planet's orbit and the angle *SPE* is a right angle. Therefore the angle of elongation, *SEP*, is the largest angle by which the inferior planet can deviate from the sun. The basic geometry of the system fully accounts for the way in which Mercury and Venus are bound to the sun.

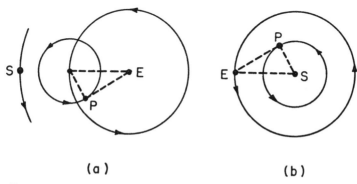

(a) **(b)**

Figure 35. Limited elongation of inferior planets explained in (*a*) the Ptolemaic and (*b*) the Copernican systems. In the Ptolemaic system the angle between the sun, S, and the planet, P, must be restricted by keeping the center of the epicycle on the line between the earth and the sun. In the Copernican system, with the planet's orbit entirely contained by the earth's, no such restriction is necessary.

Copernican geometry illuminates another even more important aspect of the behavior of the inferior planets, namely, the order of their orbits. In the Ptolemaic system the planets were arranged in earth-centered orbits so that the average distance between a planet and the earth increased with the time required for the planet to traverse the ecliptic. The device worked well for the superior planets and for the moon, but Mercury, Venus, and the sun all require 1 year for an average journey around the ecliptic, and the order of their orbits had therefore always been a source of debate. In the Copernican system there is no place for similar debate; no two planets have the same orbital period. The moon is no longer involved in the problem, for it travels about the earth rather than about the central sun. The superior planets, Mars, Jupiter, and Saturn, preserve their old order about the new center, because their orbital periods are the same as the average lengths of time they need to circle the ecliptic. The earth's orbit lies inside of Mars's, since the earth's orbital period, 1 year, is less than Mars's 687 days. It only remains to place Mercury and Venus in the system, and their order is, for the first time, uniquely determined.

This can be seen as follows. Venus is known to retrogress every 584 days, and since retrograde motion can be observed only when Venus passes the earth, 584 days must be the time Venus requires to lap the earth once in their common circuit of the sun. Now in 584 days the earth has traversed its orbit $\frac{584}{365}(=1\frac{219}{365})$ times. Since Venus has lapped the earth once during this interval, it must have circled its orbit $2\frac{219}{365}(=\frac{949}{365})$ times in just 584 days. But a planet that circles its orbit $\frac{949}{365}$ times in 584 days must require $584\times\frac{365}{949}(=225)$ days to circle its orbit once. Therefore, since Venus's period, 225 days, is less than earth's, Venus's orbit must be inside the earth's, and there is no ambiguity. A similar calculation places Mercury's orbit inside Venus's and closest to the sun. Since Mercury retrogresses, and therefore laps the earth, every 116 days, it must complete its orbit just $1\frac{116}{365}(=\frac{481}{365})$ times in 116 days. Therefore it will complete its orbit just once in $116\times\frac{365}{481}(=88)$ days. Its orbital period of 88 days is the shortest of all, and it is therefore the planet closest to the sun.

So far we have ordered the sun-centered planetary orbits with the same device used by Ptolemaic astronomers to order earth-centered orbits: planets farther from the center of the universe take longer to circle the center. The assumption that the size of the orbit increases with orbital period can be applied more fully in the Copernican than in the Ptolemaic system, but in both systems it is initially arbitrary. It seems natural that planets should behave this way, like Vitruvius' ants on a wheel, but there is no necessity that they do so. Perhaps the assumption is entirely gratuitous, and the planets, excepting the sun and moon, whose distances can be directly determined, have another order.

The response to this suggested reordering constitutes another very important difference between the Copernican and the Ptolemaic systems, and one which, as we discovered in his preface, Copernicus himself particularly emphasizes. In the Ptolemaic system the deferent and epicycle of any one planet can be shrunk or expanded at will without affecting either the sizes of the other planetary orbits or the position at which the planet, viewed from a central earth, appears against the stars. The order of the orbits *may be* determined by assuming a relation between size of orbit and orbital period. In addition, the relative dimensions of the orbits *may be* worked out with the aid of the further assumption, discussed in Chapter 3, that the minimum distance of one planet from the earth is just equal to the maximum distance between the earth and the next interior planet. But though both of these seem natural assumptions, neither is necessary. The Ptolemaic system could predict the same apparent positions for the planets without making use of either. In the Ptolemaic system the appearances are not dependent upon the order or the sizes of the planetary orbits.

There is no similar freedom in the Copernican system. If all the planets revolve in approximately circular orbits about the sun, then both the order and the relative sizes of the orbits can be determined directly from observation without additional assumptions. Any change in order or even in relative size of the orbits will upset the whole system. For example, Figure 36a shows an inferior planet, P, viewed from the earth at the time when it reaches its maximum elongation from the sun. The orbit is assumed circular, and the angle SPE must therefore be a right angle when the angle of elongation, SEP, reaches its maximum value. The planet, the sun, and the earth form a right triangle one of whose acute angles, SEP, can be directly measured. But knowledge of one acute angle of a right triangle determines the ratio of the lengths of the sides of that triangle. Therefore the ratio of the radius of the inferior planet's orbit, SP, to the radius of the earth's orbit, SE, can be computed from the measured value of the angle SEP. The relative sizes of the earth's orbit and the orbits of both inferior planets can be discovered from observation.

An equivalent determination can be made for a superior planet, though the techniques are more complex. One possible technique is illustrated in Figure 36b. Suppose that at some determined instant of time the sun, the earth, and the planet all lie on the straight line SEP; this is the orientation in which the planet lies diametrically across the ecliptic from the sun and is in the middle of a retrograde motion. Since the earth traverses its orbit more rapidly than any su-

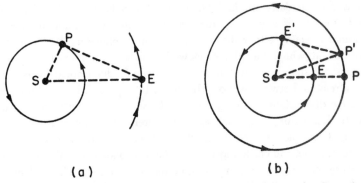

(a)   (b)

Figure 36. Determining the relative dimensions of orbits in the Copernican system: (*a*) for an inferior planet; (*b*) for a superior planet.

perior planet, there must be some later instant of time when the earth at E' and the planet at P' will form a right angle $SE'P'$ with the sun, and since $SE'P'$ is the angle between the sun and the superior planet viewed from the earth, it can be directly determined and the time required to achieve it can be measured. The angle ESE' can now be determined, for it must bear the same ratio to 360° as the time required by the earth to move from E to E' bears to the 365 days that the earth requires to complete its orbit. The angle PSP' can be determined in just the same way, since the time required by the planet to complete its orbit is already known, and the time occupied by the planet in going from P to P' is the same as that needed by the earth to go from E to E'. With PSP' and ESE' known, the angle $P'SE'$ can be found by subtraction. Then we again have a right triangle, $SE'P'$, with one acute angle, $P'SE'$, known, and the ratio of the radius of the planet's orbit, SP', to that of the earth's orbit, SE', can therefore be determined just as for an inferior planet.

By techniques like this the distances to all the planets can be determined in terms of the distance between the earth and the sun, or in terms of any unit, like the stade, in which the radius of the earth's orbit has been measured. Now, for the first time, as Copernicus says in his prefatory letter, "the orders and magnitudes of all stars and spheres . . . become so bound together that nothing in any part thereof could be moved from its place without producing confusion of all the other parts and of the universe as a whole." Because the relative dimensions of the planetary orbits are a direct consequence of the first geometric premises of sun-centered astronomy, the new astronomy has for Copernicus a naturalness and coherence that were lacking in the older earth-centered version. The structure of the heavens can be derived from Copernicus' system with fewer extraneous or *ad hoc* assumptions like plenitude. That is the new and aesthetic harmony which Copernicus emphasizes and illustrates so fully in the tenth chapter of his introductory First Book, to which we now turn, having first learned enough about the new system (as Copernicus' lay readers had not) to understand what he is talking about.

10. *Of the Order of the Heavenly Bodies.*

No one doubts that the Sphere of the Fixed Stars is the most distant of visible things. As for the order of the planets, the early Philosophers wished to determine it from the magnitude of their revolutions. They adduce the fact that of objects moving with equal speed, those farther distant seem to move more slowly (as is proved in Euclid's *Optics*). They think that the Moon describes her path in the shortest time because, being nearest to the Earth, she revolves in the smallest circle. Farthest they place Saturn, who in the longest time describes the greatest circuit. Nearer than he is Jupiter, and then Mars.

Opinions differ as to Venus and Mercury which, unlike the others, do not altogether leave the Sun. Some place them beyond the Sun, as Plato in *Timaeus*; others nearer than the Sun, as Ptolemy and many of the moderns. Alpetragius [a twelfth-century Moslem astronomer] makes Venus nearer and Mercury farther than the Sun. If we agree with Plato in thinking that the planets are themselves dark bodies that do but reflect light from the Sun, it must follow, that if nearer than the Sun, on account of their proximity to him they would appear as half or partial circles; for they would generally reflect such light as they receive upwards, that is toward the Sun, as with the waxing or waning Moon. [See the discussion of the phases of Venus in the next chapter. Neither this effect nor the following is distinctly visible without the telescope.] Some think that since no eclipse even proportional to their size is ever caused by these planets, they can never be between us and the Sun. . . . [Copernicus proceeds to note many difficulties in the arguments usually used to determine the relative order of the sun and the inferior planets. Then he continues:]

Unconvincing too is Ptolemy's proof that the Sun moves between those bodies that do and those that do not recede from him completely [that is, between the superior planets which can assume any angle of elongation and the inferior planets whose maximum elongation is limited]. Consideration of the case of the Moon, which does so recede, exposes its falseness. Again, what cause can be alleged, by those who place Venus nearer than the Sun, and Mercury next, or in some other order? Why should not these planets also follow separate paths, distinct from that of the Sun, as do the other planets [whose deferents are not tied to the sun's]? And this might be said even if their relative swiftness and slowness did not belie their alleged order. Either then the Earth cannot be the center to which the order of the planets and their Spheres is related, or certainly their relative order is not observed, nor does it appear why a higher position should be assigned to Saturn than to Jupiter, or any other planet.

Therefore I think we must seriously consider the ingenious view held by Martianus Capella [a Roman encyclopedist of the fifth century who recorded a theory of the inferior planets probably first suggested by Heraclides] . . . and certain other Latins, that Venus and Mercury do not go round the Earth like the other planets but run their courses with the Sun as center, and so do not depart from him farther than the convexity of their Spheres allows. . . . What else can they mean than that the center of these Spheres is near the Sun? So certainly the circle of Mercury must be within that of Venus, which, it is agreed, is more than twice as great.

We may now extend this hypothesis to bring Saturn, Jupiter and Mars also into relation with this center, making their Spheres great enough to contain those of Venus and Mercury and the Earth. . . . These outer planets are always nearer to the Earth about the time of their evening rising, that is, when they are in opposition to the Sun, and the Earth between them and the Sun. They are more distant from the Earth at the time of their evening setting, when they are in conjunction with the Sun and the Sun between them and the Earth. These indications prove that their center pertains rather to the Sun than to the Earth, and that this is the same center as that to which the revolutions of Venus and Mercury are related.

[Copernicus' remarks do not actually "prove" a thing. The Ptolemaic system explains these phenomena as completely as the Copernican, but the Copernican explanation is again more natural, for, like the Copernican explanation of the limited elongation of the inferior planets, it depends only on the geometry of a sun-centered astronomical system, not on the particular orbital periods assigned to the planets. Copernicus' remarks will be clarified by reference to Figure 32a. A superior planet retrogresses when the earth overtakes it, and under these circumstances it must be simultaneously closest to the earth and across the ecliptic from the sun. In the Ptolemaic system a retrogressing superior planet must be closer to the earth than at any other time, and it is in fact also across the sky from the sun. But it is only across the sky from the sun because the rates of rotation of its deferent and epicycle have particular values that happen to put the planet back in opposition to the sun whenever the epicycle brings the planet back close to the central earth. If, in the Ptolemaic system, the period of epicycle or deferent were quantitatively slightly different, then the qualitative regularity that puts a retrogressing superior planet across the sky from the sun would not occur. In the Copernican system it must occur regardless of the particular rates at which the planets revolve in their orbits.]

But since all these [Spheres] have one center it is necessary that the space between the convex side of Venus's Sphere and the concave side of Mars's must also be viewed as a Sphere concentric with the others, capable of receiving the Earth with her satellite the Moon and whatever is contained within the Sphere of the Moon — for we must not separate the Moon from the Earth, the former being beyond all doubt nearest to the latter, especially as in that space we find suitable and ample room for the Moon.

We therefore assert that the center of the Earth, carrying the Moon's path, passes in a great circuit among the other planets in an annual revolution round the Sun; that near the Sun is the center of the Universe; and that whereas the Sun is at rest, any apparent motion of the Sun can be better explained by motion of the Earth. Yet so great is the Universe that though the distance of the Earth from the Sun is not insignificant compared with the size of any other planetary path, in accordance with the ratios of their sizes, it is insignificant compared with the distances of the Sphere of the Fixed Stars.

I think it easier to believe this than to confuse the issue by assuming a vast number of Spheres, which those who keep Earth at the center must do. We thus rather follow Nature, who producing nothing vain or superfluous often prefers to endow one cause with many effects. Though these views are difficult, contrary to expectation, and certainly unusual, yet in the sequel we shall, God willing, make them abundantly clear at least to mathematicians.

Given the above view — and there is none more reasonable — that the periodic times are proportional to the sizes of the Spheres, then the order of the Spheres, beginning from the most distant is as follows. Most distant of all is the Sphere of the Fixed Stars, containing all things, and being therefore itself immovable. It represents that to which the motion and position of all the other bodies must be referred . . . . Next is the planet Saturn, revolving in 30 years. Next comes Jupiter, moving in a 12-year circuit; then Mars, who goes round in 2 years. The fourth place is held by the annual revolution [of the Sphere] in which the Earth is contained, together with the Sphere of the Moon as on an epicycle. Venus, whose period is 9 months, is in the fifth place, and sixth is Mercury, who goes round in the space of 80 days.

In the middle of all sits Sun enthroned. In this most beautiful temple could we place this luminary in any better position from which he can illuminate the whole at once? He is rightly called the Lamp, the Mind, the

Ruler of the Universe; Hermes Trismegistus names him the Visible God, Sophocles' Electra calls him the All-seeing. So the Sun sits as upon a royal throne ruling his children the planets which circle round him. The Earth has the Moon at her service. As Aristotle says, in his On [the Generation of] Animals, the Moon has the closest relationship with the Earth. Meanwhile the Earth conceives by the Sun, and becomes pregnant with an annual rebirth.

So we find underlying this ordination an admirable symmetry in the Universe, and a clear bond of harmony in the motion and magnitude of the Spheres such as can be discovered in no other wise. For here we may observe why the progression and retrogression appear greater for Jupiter than Saturn, and less than for Mars, but again greater for Venus than for Mercury [a glance at Figure 32 will show that the closer the orbit of a planet is to the orbit of the earth, the larger the apparent retrograde motion of that planet must be – an additional harmony of Copernicus' system]; and why such oscillation appears more frequently in Saturn than in Jupiter, but less frequently in Mars and Venus than in Mercury [the earth will lap a slowly moving superior planet more frequently than it laps a rapid one, and conversely for an inferior planet]; moreover why Saturn, Jupiter and Mars are nearer to the Earth at opposition to the Sun than when they are lost in or emerge from the Sun's rays. Particularly Mars, when he shines all night [and is therefore in opposition], appears to rival Jupiter in magnitude, being only distinguishable by his ruddy color; otherwise he is scarce equal to a star of the second magnitude, and can be recognized only when his movements are carefully followed. All these phenomena proceed from the same cause, namely Earth's motion.

That there are no such phenomena for the fixed stars proves their immeasurable distance, because of which the outer sphere's [apparent] annual motion or its [parallactic] image is invisible to the eyes. For every visible object has a certain distance beyond which it can no more be seen, as is proved in optics. The twinkling of the stars, also, shows that there is still a vast distance between the farthest of the planets, Saturn, and the Sphere of the Fixed Stars [for if the stars were very near Saturn, they should shine as he does], and it is chiefly by this indication that they are distinguished from the planets. Further, there must necessarily be a great difference between moving and non-moving bodies. So great is this divine work of the Great and Noble Creator!

Throughout this crucially important tenth chapter Copernicus' emphasis is upon the "admirable symmetry" and the "clear bond of harmony in the motion and magnitude of the Spheres" that a sun-centered geometry imparts to the appearances of the heavens. If the sun is the center, then an inferior planet cannot possibly appear far from the sun; if the sun is the center, then a superior planet must be in opposition to the sun when it is closest to the earth; and so on and on. It is through arguments like these that Copernicus seeks to persuade his contemporaries of the validity of his new approach. Each argument cites an aspect of the appearances that can be explained by *either* the Ptolemaic *or* the Copernican system, and each then proceeds to point out how much more harmonious, coherent, and natural the Copernican explanation is. There are a great many such arguments. The sum of the evidence drawn from harmony is nothing if not impressive.

But it may well be nothing. "Harmony" seems a strange basis on which to argue for the earth's motion, particularly since the harmony is so obscured by the complex multitude of circles that make up the full Copernican system. Copernicus' arguments are not pragmatic. They appeal, if at all, not to the utilitarian sense of the practicing astronomer but to his aesthetic sense and to that alone. They had no appeal to laymen, who, even when they understood the arguments, were unwilling to substitute minor celestial harmonies for major terrestrial discord. They did not necessarily appeal to astronomers, for the harmonies to which Copernicus' arguments pointed did not enable the astronomer to perform his job better. New harmonies did not increase accuracy or simplicity. Therefore they could and did appeal primarily to that limited and perhaps irrational subgroup of mathematical astronomers whose Neoplatonic ear for mathematical harmonies could not be obstructed by page after page of complex mathematics leading finally to numerical predictions scarcely better than those they had known before. Fortunately, as we shall discover in the next chapter, there were a few such astronomers. Their work is also an essential ingredient of the Copernican Revolution.

Revolution by Degrees

Because he was the first fully to develop an astronomical system based upon the motion of the earth, Copernicus is frequently called the first modern astronomer. But, as the text of the *De Revolutionibus* indicates, an equally persuasive case might be made for calling him the last great Ptolemaic astronomer. Ptolemaic astronomy meant far more than astronomy predicated on a stationary earth, and it is only with respect to the position and motion of the earth that Copernicus broke with the Ptolemaic tradition. The cosmological frame in which his astronomy was embedded, his physics, terrestrial and celestial, and even the mathematical devices that he employed to make his system give adequate predictions are all in the tradition established by ancient and medieval scientists.

Though historians have occasionally grown livid arguing whether Copernicus is really the last of the ancient or the first of the modern astronomers, the debate is in principle absurd. Copernicus is neither an ancient nor a modern but rather a Renaissance astronomer in whose work the two traditions merge. To ask whether his work is really ancient or modern is rather like asking whether the bend in an otherwise straight road belongs to the section of road that precedes the bend or to the portion that comes after it. From the bend both sections of the road are visible, and its continuity is apparent. But viewed from a point before the bend, the road seems to run straight to the bend and then to disappear; the bend seems the last point in a straight road. And viewed from a point in the next section, after the bend, the road appears to begin at the bend from which it runs straight on. The bend belongs equally to both sections, or it belongs to neither. It marks a turning point in the direction of the road's progress, just as the *De Revolutionibus* marks a shift in the direction in which astronomical thought developed.

To this point in this chapter we have emphasized primarily the ties between the *De Revolutionibus* and the earlier astronomical and cosmological tradition. We have minimized, as Copernicus himself does, the extent of the Copernican innovation, because we have been concerned to discover how a potentially destructive innovation could be produced by the tradition that it was ultimately to destroy. But, as we shall soon discover, this is not the only legitimate way to view the *De Revolutionibus*, and it is not the view taken by most later Copernicans. For Copernicus' sixteenth- and seventeenth-century followers, the primary importance of the *De Revolutionibus* derived from its single novel concept, the planetary earth, and from the novel astronomical consequences, the new harmonies, which Copernicus had derived from that concept. To them Copernicanism meant the threefold motion of the earth and, initially, that alone. The traditional conceptions with which Copernicus had clothed his innovation were not to his followers essential elements of his work, simply because, as traditional elements, they were not Copernicus' contribution to science. It was not because of its traditional elements that people quarreled about the *De Revolutionibus*.

That is why the *De Revolutionibus* could be the starting point for a new astronomical and cosmological tradition as well as the culmination of an old one. Those whom Copernicus converted to the concept of a moving earth began their research from the point at which Copernicus had stopped. Their starting point was the earth's motion, which was all they necessarily took from Copernicus, and the problems to which they devoted themselves were not the problems of the old astronomy, which had occupied Copernicus, but the problems of the new sun-centered astronomy, which they discovered in the *De Revolutionibus*. Copernicus presented them with a set of problems that neither he nor his predecessors had had to face. In the pursuit of those problems the Copernican Revolution was completed, and a new astronomical tradition, deriving from the *De Revolutionibus*, was founded. Modern astronomy looks back to the *De Revolutionibus* as Copernicus had looked back to Hipparchus and Ptolemy.

Major upheavals in the fundamental concepts of science occur by degrees. The work of a single individual may play a preëminent role in such a conceptual revolution, but if it does, it achieves preëminence either because, like the *De Revolutionibus*, it initiates revolution by a small innovation which presents science with new problems, or because, like Newton's *Principia*, it terminates revolution by integrating concepts derived from many sources. The extent of the innovation that any individual can produce is necessarily limited, for each individual must employ in his research the tools that he acquires from a traditional education, and he cannot in his own lifetime replace them all. It seems therefore that many of the elements in the *De Revolutionibus* which, in the earlier parts of this chapter, we pointed to as incongruities are not really incongruities at all. The *De Revolutionibus* seems incongruous only to those who expect to find the entire Copernican Revolution in the work which gives that revolution its name, and such an expectation derives from a misunderstanding of the way in which new patterns of scientific thought are produced. The limitations of the *De Revolutionibus* might better be regarded as essential and typical characteristics of any revolution-making work.

Most of the apparent incongruities in the *De Revolutionibus* reflect the personality of its author, and Copernicus' personality seems entirely appropriate to his seminal role in the development of astronomy. Copernicus was a dedicated specialist. He belonged to the revived Hellenistic tradition of mathematical astronomy which emphasized the mathematical problem of the planets at the expense of cosmology. For his Hellenistic predecessors the physical incongruity of an epicycle had not been an important drawback of the Ptolemaic system, and Copernicus displayed a similar indifference to cosmological detail when he failed to note the incongruities of a moving earth in an otherwise traditional universe. For him, mathematical and celestial detail came first; he wore blinders that kept his gaze focused upon the mathematical harmonies of the heavens. To anyone who did not share his specialty Copernicus' view of the universe was narrow and his sense of values distorted.

But an excessive concern with the heavens and a distorted sense of values may be essential characteristics of the man who inaugurated the revolution in astronomy and cosmology. The blinders that restricted Copernicus' gaze to the heavens may have been functional. They made him so perturbed by discrepancies of a few degrees in astronomical prediction that in an attempt to resolve them he could embrace a cosmological heresy, the earth's motion. They gave him an eye so absorbed with geometrical harmony that he could adhere to his heresy for its harmony alone, even when it had failed to solve the problem that had led him to it. And they helped him evade the nonastronomical consequences of his innovation, consequences that led men of less restricted vision to reject his innovation as absurd.

Above all, Copernicus' dedication to the celestial motions is responsible for the painstaking detail with which he explored the mathematical consequences of the earth's motion and fitted those consequences to an existing knowledge of the heavens. That detailed technical study is Copernicus' real contribution. Both before and after Copernicus there were cosmologists more radical than he, men who with broad brush strokes sketched an infinite and multipopulated universe. But none of them produced work resembling the later books of the *De Revolutionibus*, and it is these books which, by showing for the first time that the astronomer's job could be done, and done more harmoniously, from a moving earth, provided a stable base from which to launch a new astronomical tradition. Had Copernicus' cosmological First Book appeared alone, the Copernican Revolution would and should be known by someone else's name.

6

THE ASSIMILATION OF

COPERNICAN ASTRONOMY

The Reception of Copernicus' Work

Copernicus died in 1543, the year in which the *De Revolutionibus* was published, and tradition tells us that he received the first printed copy of his life's work on his deathbed. The book had to fight its battles without further help from its author. But for those battles Copernicus had constructed an almost ideal weapon. He had made the book unreadable to all but the erudite astronomers of his day. Outside of the astronomical world the *De Revolutionibus* created initially very little stir. By the time large-scale lay and clerical opposition developed, most of the best European astronomers, to whom the book was directed, had found one or another of Copernicus' mathematical techniques indispensable. It was then impossible to suppress the work completely, particularly because it was in a printed book and not, like Oresme's work or Buridan's, in a manuscript. Whether intentionally or not, the final victory of the *De Revolutionibus* was achieved by infiltration.

For two decades before the publication of his principal work Copernicus had been widely recognized as one of Europe's leading astronomers. Reports about his research, including his new hypothesis, had circulated since about 1515. The publication of the *De Revolutionibus* was eagerly awaited. When it appeared, Copernicus' contemporaries may have been skeptical of its main hypothesis and disappointed in the complexity of its astronomical theory, but they were nevertheless forced to recognize Copernicus' book as the first European astronomical text that could rival the *Almagest* in depth and completeness. Many advanced astronomical texts written during the fifty years after Copernicus' death referred to him as a "second

Ptolemy" or "the outstanding artificer of our age"; increasingly these books borrowed data, computations, and diagrams from the *De Revolutionibus*, at least from parts of it independent of the motion of the earth. During the second half of the sixteenth century the book became a standard reference for all those concerned with advanced problems of astronomical research.

But the success of the *De Revolutionibus* does not imply the success of its central thesis. The faith of most astronomers in the earth's stability was at first unshaken. Authors who applauded Copernicus' erudition, borrowed his diagrams, or quoted his determination of the distance from the earth to the moon, usually either ignored the earth's motion or dismissed it as absurd. Even the rare text that mentioned Copernicus' hypothesis with respect rarely defended or used it. With a few notable exceptions, the most favorable of the early reactions to the Copernican innovation are typified by the remark of the English astronomer Thomas Blundeville, who wrote: "Copernicus . . . affirmeth that the earth turneth about and that the sun standeth still in the midst of the heavens, by help of which false supposition he hath made truer demonstrations of the motions and revolutions of the celestial spheres, than ever were made before." [1] Blundeville's remark appeared in 1594 in an elementary book on astronomy that took the earth's stability for granted. Yet the tenor of Blundeville's rejection must have sent his more alert and proficient readers straight to the *De Revolutionibus*, a book which, in any case, no proficient astronomer could ignore. From the start the *De Revolutionibus* was widely read, but it was read in spite of, rather than because of, its strange cosmological hypothesis.

Nevertheless, the book's large audience ensured it a small but increasing number of readers equipped to discover Copernicus' harmonies and willing to admit them as evidence. There were a few converts, and their work helped in varied ways to spread knowledge of Copernicus' system. The *Narratio Prima* or *First Account* by Copernicus' earliest disciple, George Joachim Rheticus (1514–1576), remained the best brief technical description of the new astronomical methods for many years after its first publication in 1540. The popular elementary defense of Copernicanism published in 1576 by the English astronomer Thomas Digges (c.1546–1595) did much to spread the concept of the earth's motion beyond the narrow circle of astronomers.

And the teaching and research of Michael Maestlin (1550–1631), professor of astronomy at the University of Tübingen, gained a few converts, including Kepler, for the new astronomy. Through the teaching, writing, and research of men like these, Copernicanism inevitably gained ground, though the astronomers who avowed their adherence to the conception of a moving earth remained a small minority.

But the size of the group of avowed Copernicans is not an adequate index of the success of Copernicus' innovation. Many astronomers found it possible to exploit Copernicus' mathematical system and to contribute to the success of the new astronomy while denying or remaining silent about the motion of the earth. Hellenistic astronomy provided their precedent. Ptolemy himself had never pretended that all of the circles used in the *Almagest* to compute planetary position were physically real; they were useful mathematical devices and they did not have to be any more than that. Similarly, Renaissance astronomers were at liberty to treat the circle representing the earth's orbit as a mathematical fiction, useful for computations alone; they could and occasionally did compute planetary position *as if* the earth moved without committing themselves to the physical reality of that motion. Andreas Osiander, the Lutheran theologian who saw Copernicus' manuscript through the press, had actually urged this alternative upon readers in an anonymous preface attached to the *De Revolutionibus* without Copernicus' permission. The spurious preface probably did not fool many astronomers, but a number of them nevertheless took advantage of the alternative that it suggested. Using Copernicus' mathematical system without advocating the physical motion of the earth provided a convenient escape from the dilemma posed by the contrasting celestial harmonies and terrestrial discord of the *De Revolutionibus*. It also gradually tempered the astronomer's initial conviction that the earth's motion was absurd.

Erasmus Reinhold (1511–1553) was the first astronomer to do important service for the Copernicans without declaring himself in favor of the earth's motion. In 1551, only eight years after the publication of the *De Revolutionibus*, he issued a complete new set of astronomical tables, computed by the mathematical methods developed by Copernicus, and these soon became indispensable to astronomers and astrologers, whatever their beliefs about the position and motion of the earth. Reinhold's *Prutenic Tables*, named for his patron, the

Duke of Prussia, were the first complete tables prepared in Europe for three centuries, and the old tables, which had included some errors from the start, were now badly out of date — the clock had run too long. Reinhold's supremely careful work, based on somewhat more and better data than had been available to the men who computed the thirteenth-century tables, produced a set of tables which, for most applications, were measurably superior to the old. They were not, of course, completely accurate; Copernicus' mathematical system was intrinsically no more accurate than Ptolemy's; errors of a day in the prediction of lunar eclipses were common, and the length of the year determined from the *Prutenic Tables* was actually slightly less accurate than that determined from the older tables. But most comparisons displayed the superiority of Reinhold's work, and his tables became increasingly an astronomical requisite. Since the tables were known to derive from the astronomical theory of the *De Revolutionibus*, Copernicus' prestige inevitably gained. Every man who used the *Prutenic Tables* was at least acquiescing in an implicit Copernicanism.

During the second half of the sixteenth century astronomers could dispense with neither the *De Revolutionibus* nor the tables based upon it. Copernicus' proposal gained ground slowly but apparently inexorably. Successive generations of astronomers, decreasingly predisposed by experience and training to take the earth's stability for granted, found the new harmonies a more and more forceful argument for its motion. Besides, by the end of the century the first converts had begun to uncover new evidence. Therefore if the decision between the Copernican and the traditional universe had concerned only astronomers, Copernicus' proposal would almost certainly have achieved a quiet and gradual victory. But the decision was not exclusively, or even primarily, a matter for astronomers, and as the debate spread from astronomical circles it became tumultuous in the extreme. To most of those who were not concerned with the detailed study of celestial motions, Copernicus' innovation seemed absurd and impious. Even when understood, the vaunted harmonies seemed no evidence at ll. The resulting clamor was widespread, vocal, and bitter.

But the clamor was slow in starting. Initially, few nonastronomers knew of Copernicus' innovation or recognized it as more than a passing individual aberration like many that had come and gone before. Most of the elementary astronomy texts and manuals used during the second

half of the sixteenth century had been prepared long before Copernicus' lifetime — John of Holywood's thirteenth-century primer was still
a leader in elementary training — and the new handbooks prepared
after the publication of the *De Revolutionibus* usually did not mention
Copernicus or dismissed his innovation in a sentence or two. The
popular cosmological books that described the universe to laymen
remained even more exclusively Aristotelian in tone and substance;
Copernicus was either unknown to their authors or, if known, he was
usually ignored. Except, perhaps, in a few centers of Protestant learning, Copernicanism does not seem to have been a cosmological issue
during the first few decades after Copernicus' death. Outside of astronomical circles it seldom became a major issue until the beginning of
the seventeenth century.

There were a few sixteenth-century reactions from nonastronomers,
and they provide a foretaste of the immense debate to follow, for they
were usually unequivocally negative. Copernicus and his few followers
were ridiculed for the absurdity of their concept of a moving earth,
though without the bitterness or the elaborate dialectic which developed when it became apparent that Copernicanism was to be a
stubborn and dangerous opponent. One long cosmological poem, first
published in France in 1578 and immensely popular there and in
England during the next century and a quarter, provides the following typical description of the Copernicans as

> Those clerks who think (think how absurd a jest)
> That neither heav'ns nor stars do turn at all,
> Nor dance about this great round earthly ball;
> But th'earth itself, this massy globe of ours,
> Turns round-about once every twice-twelve hours:
> And we resemble land-bred novices
> New brought aboard to venture on the seas;
> Who, at first launching from the shore, suppose
> The ship stands still, and that the ground it goes. . . .
>   So, never should an arrow, shot upright,
> In the same place upon the shooter light;
> But would do, rather, as, at sea, a stone
> Aboard a ship upward uprightly thrown;
> Which not within-board falls, but in the flood
> Astern the ship, if so the wind be good.
> So should the fowls that take their nimble flight
> From western marches towards morning's light; . . .

And bullets thundered from the cannon's throat
(Whose roaring drowns the heav'nly thunder's note)
Should seem recoil: since the quick career,
That our round earth should daily gallop here,
Must needs exceed a hundred-fold, for swift,
Birds, bullets, winds; their wings, their force, their drift.
    Arm'd with these reasons, 'twere superfluous
T'assail the reasons of Copernicus;
Who, to save better of the stars th'appearance,
Unto the earth a three-fold motion warrants.[2]

Since the author of this poetic rejection of Copernicanism was a poet, not a scientist or philosopher, his cosmological conservatism and his adherence to classic sources may not be surprising. Yet it was from poets and popularizers rather than from astronomers that most people in the sixteenth and seventeenth century, as today, learned about the universe. Du Bartas's *The Week, or the Creation of the World,* from which the preceding excerpt is taken, was a far more widely read and influential book than the *De Revolutionibus.*

In any case, uncritical offhand condemnations of Copernicus and his followers were not restricted to conservative and unoriginal popularizers. Jean Bodin, famous as one of the most advanced and creative political philosophers of the sixteenth century, discards Copernicus' innovation in almost identical terms:

No one in his senses, or imbued with the slightest knowledge of physics, will ever think that the earth, heavy and unwieldy from its own weight and mass, staggers up and down around its own center and that of the sun; for at the slightest jar of the earth, we would see cities and fortresses, towns and mountains thrown down. A certain courtier Aulicus, when some astrologer in court was upholding Copernicus' idea before Duke Albert of Prussia, turning to the servant who was pouring the Falernian, said: "Take care that the flagon is not spilled." For if the earth were to be moved, neither an arrow shot straight up, nor a stone dropped from the top of a tower would fall perpendicularly, but either ahead or behind. . . . Lastly, all things on finding places suitable to their natures, remain there, as Aristotle writes. Since therefore the earth has been allotted a place fitting its nature, it cannot be whirled around by other motion than its own.[3]

In this passage Bodin looks a traditionalist, but he was not. Because of its generally radical and atheistic tone, the book from which the quotation is taken was in 1628 placed upon the Index of books that Catholics are forbidden to read. Although its author was himself a

Catholic, the book remains there to this day. Bodin was quite willing to break with tradition, but that was not enough to make a man a Copernican. It was almost invariably also necessary to understand astronomy and to take its problems immensely seriously. Except to those with an astronomical bias, the earth's motion seemed very nearly as absurd in the years after Copernicus' death as it had before.

The anti-Copernican arguments suggested by Du Bartas and Bodin can be considerably elaborated along lines anticipated by our discussions of the Aristotelian universe in Chapters 3 and 4. In one or another disguise, which we need not penetrate, they appear again and again during the first half of the seventeenth century when the debate about the earth's motion became bitter and intense. The earth's motion, it was said, violates the first dictate of common sense; it conflicts with long-established laws of motion; it has been suggested merely "to save better of the stars th'appearance," a ridiculously minuscule incentive for revolution. These are forceful arguments, quite sufficient to convince most people. But they are not the most forceful weapons in the anti-Copernican battery, and they are not the ones that generated the most heat. Those weapons were religious and, particularly, scriptural.

Citation of Scripture against Copernicus began even before the publication of the *De Revolutionibus*. In one of his "Table Talks," held in 1539, Martin Luther is quoted as saying:

People gave ear to an upstart astrologer who strove to show that the earth revolves, not the heavens or the firmament, the sun and the moon. . . . This fool wishes to reverse the entire science of astronomy; but sacred Scripture tells us [Joshua 10:13] that Joshua commanded the sun to stand still, and not the earth.[4]

Luther's principal lieutenant, Melanchthon, soon joined in the increasing Protestant clamor against Copernicus. Six years after Copernicus' death he wrote:

The eyes are witnesses that the heavens revolve in the space of twenty-four hours. But certain men, either from the love of novelty, or to make a display of ingenuity, have concluded that the earth moves; and they maintain that neither the eighth sphere nor the sun revolves. . . . Now, it is a want of honesty and decency to assert such notions publicly, and the example is pernicious. It is the part of a good mind to accept the truth as revealed by God and to acquiesce in it.[5]

Melanchthon then proceeded to assemble a number of anti-Copernican Biblical passages, emphasizing the famous verses, Ecclesiastes 1:4–5, which state "the earth abideth forever" and that "The sun also ariseth, and the sun goeth down, and hasteth to his place where he arose." Finally he suggests that severe measures be taken to restrain the impiety of the Copernicans.

Other Protestant leaders soon joined in the rejection of Copernicus. Calvin, in his *Commentary on Genesis*, cited the opening verse of the Ninety-third Psalm — "the earth also is stablished, that it cannot be moved" — and he demanded, "Who will venture to place the authority of Copernicus above that of the Holy Spirit?" [6] Increasingly, Biblical citation became a favored source of anti-Copernican argument. By the first decades of the seventeenth century clergymen of many persuasions were to be found searching the Bible line by line for a new passage that would confound the adherents of the earth's motion. With growing frequency Copernicans were labeled "infidel" and "atheist," and when, after about 1610, the Catholic Church officially joined the battle against Copernicanism, the charge became formal heresy. In 1616 the *De Revolutionibus* and all other writings that affirmed the earth's motion were put upon the Index. Catholics were forbidden to teach or even to read Copernican doctrines, except in versions emended to omit all reference to the moving earth and central sun.

The preceding sketch displays the most popular and forceful weapons in the arsenal arrayed against Copernicus and his followers, but it scarcely indicates what the war was really about. Most of the men quoted above are so ready to reject the earth's motion as absurd or as conflicting with authority that they fail to show, and may not at first have realized fully, that Copernicanism was potentially destructive of an entire fabric of thought. Their very dogmatism disguises their motives. But it does not eliminate them. More than a picture of the universe and more than a few lines of Scripture were at stake. The drama of Christian life and the morality that had been made dependent upon it would not readily adapt to a universe in which the earth was just one of a number of planets. Cosmology, morality, and theology had long been interwoven in the traditional fabric of Christian thought described by Dante at the beginning of the fourteenth century. The vigor and venom displayed at the height of the Copernican controversy, three centuries later, testifies to the strength and vitality of the tradition.

When it was taken seriously, Copernicus' proposal raised many gigantic problems for the believing Christian. If, for example, the earth were merely one of six planets, how were the stories of the Fall and of the Salvation, with their immense bearing on Christian life, to be preserved? If there were other bodies essentially like the earth, God's goodness would surely necessitate that they, too, be inhabited. But if there were men on other planets, how could they be descendants of Adam and Eve, and how could they have inherited the original sin, which explains man's otherwise incomprehensible travail on an earth made for him by a good and omnipotent deity? Again, how could men on other planets know of the Saviour who opened to them the possibility of eternal life? Or, if the earth is a planet and therefore a celestial body located away from the center of the universe, what becomes of man's intermediate but focal position between the devils and the angels? If the earth, as a planet, participates in the nature of celestial bodies, it can not be a sink of iniquity from which man will long to escape to the divine purity of the heavens. Nor can the heavens be a suitable abode for God if they participate in the evils and imperfection so clearly visible on a planetary earth. Worst of all, if the universe is infinite, as many of the later Copernicans thought, where can God's Throne be located? In an infinite universe, how is man to find God or God man?

These questions have answers. But the answers were not easily achieved; they were not inconsequential; and they helped to alter the religious experience of the common man. Copernicanism required a transformation in man's view of his relation to God and of the bases of his morality. Such a transformation could not be worked out overnight, and it was scarcely even begun while the evidence for Copernicanism remained as indecisive as it had been in the *De Revolutionibus*. Until that transformation was achieved, sensitive observers might well find traditional values incompatible with the new cosmology, and the frequency with which the charge of atheism was hurled at the Copernicans is evidence of the threat to the established order posed to many observers by the concept of a planetary earth.

But the charge of atheism is only indirect evidence. More forceful testimony comes from men who felt compelled to take the Copernican

innovation seriously. As early as 1611, the English poet and divine
John Donne said to the Copernicans that "those opinions of yours may
very well be true. . . . [In any case, they are now] creeping into
every man's mind," [7] but he could discover little except evil in the
impending transition. During the same year in which he reluctantly
conceded the probability of the earth's motion, he portrayed his discomfort at the impending dissolution of traditional cosmology in *The
Anatomy of the World*, a poem in which "the frailty and decay of this
whole world is represented." Part of Donne's malaise derived specifi-
cally from Copernicanism:

> [The] new Philosophy calls all in doubt,
> The Element of fire is quite put out;
> The Sun is lost, and th'earth, and no man's wit
> Can well direct him where to look for it.
> And freely men confess that this world's spent,
> When in the Planets, and the Firmament
> They seek so many new; then see that this
> Is crumbled out again to his Atomies.
> 'Tis all in pieces, all coherence gone;
> All just supply, and all Relation:
> Prince, Subject, Father, Son, are things forgot,
> For every man alone thinks he hath got
> To be a Phoenix, and that then can be
> None of that kind, of which he is, but he.[8]

Fifty-six years later, when scientists, at least, had overwhelmingly
accepted the earth's motion and its status as a planet, Copernicanism
presented the same problem of Christian morality to the English poet
John Milton, though he resolved it differently. Milton, like Donne,
thought that Copernicus' innovation might very well be true. He included in *Paradise Lost* a lengthy description of the two opposing
systems of the world, the Ptolemaic and the Copernican, and he refused to take sides in what he described as the abstruse technical
controversy between them. But in his epic, whose object was "to
justify the ways of God to man," [9] he was compelled to use a traditional cosmological frame. The universe of *Paradise Lost* is not quite
Dante's universe; Milton derives the positions of heaven and hell
from a tradition even older than Dante's. But the terrestrial stage upon
which Milton portrays man's fall is still necessarily a unique, stable,
and centrally located body, created by God for man. Though more

than a century had passed since the publication of the *De Revolutioni-*
bus, the Christian drama and the morality that had been made dependent upon it could not be adapted to a universe in which the
earth was a planet and in which new worlds could continually be
discovered "in the Planets and the Firmament."

Donne's uneasiness and Milton's cosmological choice illustrate the
extrascientific issues which, during the seventeenth century, were integral parts of the controversy over Copernicanism. These issues, even
more than its apparent absurdity or its conflict with established laws
of motion, account for the hostility that Copernicus' proposal encountered outside of scientific circles. But they may not quite account either
for the intensity of that hostility or for the willingness of both Protestant and Catholic leaders to make anti-Copernicanism an official
Church doctrine which could justify the persecution of Copernicans.
It is easy to understand the existence of strong resistance to Copernicus' innovation — its patent absurdity and destructiveness were not
offset by effective evidence — but it is difficult to understand the extreme forms which that resistance occasionally took. Before the middle
of the sixteenth century the history of Christianity offers few prece-
dents for the rigidity with which the official leaders of major religious
groups applied the literal text of Scripture to suppress a scientific and
cosmological theory. Even during the early centuries of the Catholic
Church, when distinguished Church Fathers like Lactantius had employed the Scriptures to destroy classical cosmology, there had been
no official Catholic cosmological position to which communicants
were required to adhere.

The bitterness of official Protestant opposition is, in practice, far
easier to understand than its Catholic counterpart, because the Protestants' opposition can be plausibly related to a more fundamental controversy which arose in the split between the Churches. Luther and
Calvin and their followers wished to return to a pristine Christianity,
as it could be discovered in the words of Jesus and the early Fathers
of the Church. To Protestant leaders the Bible was the single fundamental source of Christian knowledge. They vehemently rejected the
ritual and the dialectic subtleties that successive authoritarian Church
Councils had interposed between the believer and the fountainhead
of his belief. They abhorred the elaborate metaphorical and allegorical interpretation of Scripture, and their literal adherence to the Bible

in matters of cosmology had no parallel since the days of Lactantius, Basil, and Kosmas. To them Copernicus may well have seemed a symbol of all the tortuous reinterpretations which, during the later Middle Ages, had separated Christians from the basis of their belief. Therefore the violence of the thunder that official Protestantism directed at Copernicus seems almost natural. Toleration of Copernicanism would have been toleration of the very attitude toward Holy Writ and toward knowledge in general which, according to Protestants, had led Christianity astray.

Copernicanism was thus indirectly involved in the larger religious battle between the Protestant and Catholic Churches, and that involvement must account for some of the excessive bitterness the Copernican controversy evoked. Protestant leaders like Luther, Calvin, and Melanchthon led in citing Scripture against Copernicus and in urging the repression of Copernicans. Since the Protestants never possessed the police apparatus available to the Catholic Church, their repressive measures were seldom so effective as those taken later by the Catholics, and they were more readily abandoned when the evidence for Copernicanism became overwhelming. But Protestants nevertheless provided the first effective institutionalized opposition. Reinhold's silence about the physical validity of the mathematical system that he had employed in computing the *Prutenic Tables* is usually interpreted as an index of the official opposition to Copernicanism at the Protestant university of Wittenberg. Osiander, who added the spurious apologia to the beginning of the *De Revolutionibus*, was also a Protestant. Rheticus, the first outspoken defender of Copernicus' astronomy, was a Protestant, too, but his *Narratio Prima* was written while he was away from Wittenberg and before the *De Revolutionibus* appeared; after his return to Wittenberg he published no more Copernican tracts.

For sixty years after Copernicus' death there was little Catholic counterpart for the Protestant opposition to Copernicanism. Individual Catholic clergymen expressed their incredulity or abhorrence of the new conception of the earth, but the Church itself was silent. The *De Revolutionibus* was read and at least occasionally taught at leading Catholic universities. Reinhold's *Prutenic Tables*, based on Copernicus' mathematical system, were used in the reformation of the calendar promulgated for the Catholic world in 1582 by Gregory

XIII. Copernicus himself had been a cleric and a reputable one, whose judgment was widely sought on astronomical and other matters. His book was dedicated to the Pope, and among the friends who urged him to publish it were a Catholic bishop and a cardinal. During the fourteenth, fifteenth, and sixteenth centuries the Church had not imposed cosmological conformity on its members. The *De Revolutionibus* was itself a product of the latitude allowed to Churchmen in matters of science and secular philosophy, and before the *De Revolutionibus* the Church had spawned even more revolutionary cosmological concepts without theological convulsions. In the fifteenth century the eminent cardinal and papal legate Nicholas of Cusa had propounded a radical Neoplatonic cosmology and had not even bothered about the conflict between his views and Scripture. Though he portrayed the earth as a moving star, like the sun and the other stars, and though his works were widely read and had great influence, he was not condemned or even criticized by his Church.

Therefore, when in 1616, and more explicitly in 1633, the Church prohibited teaching or believing that the sun was at the center of the universe and that the earth moved around it, the Church was reversing a position that had been implicit in Catholic practice for centuries. The reversal shocked a number of devout Catholics, because it committed the Church to opposing a physical doctrine for which new evidence was being discovered almost daily, and because there clearly had been an alternative attitude open to the Church. The same devices which, in the twelfth and thirteenth centuries, had permitted the Church to embrace Ptolemy and Aristotle might, in the seventeenth century, have been applied to Copernicus' proposal. In a limited fashion they had already been applied. Oresme's fourteenth-century discussion of the earth's diurnal rotation had not ignored the scriptural evidence for the earth's immobility. He had cited two of the Biblical passages noted above and had then replied:

To the . . . argument concerning the Holy Scripture which says that the sun revolves, etc., one would say that it is here conforming to the manner of common human speech, just as is done in several [other] places, e.g., where it is written that God is repentant and that he is angry and pacified and all other things which are not just as they sound. Also appropriate to our question, we read that God covers the heaven with clouds: . . . and yet in reality the heaven covers the clouds.[10]

Though the reinterpretation demanded by Copernicanism would have been more drastic and more costly, the same sort of arguments would have sufficed. During the eighteenth and nineteenth centuries similar arguments were employed, and even in the seventeenth century, at the time when the official decision to prohibit Copernicanism was being taken, a few Catholic leaders recognized that some such far-reaching reformulation might conceivably be required. In 1615 Cardinal Bellarmine, the leader of the Church officials who one year later condemned Copernican views, wrote to the Copernican Foscarini:

> If there were a real proof that the sun is in the center of the universe, that the earth is in the third heaven, and that the sun does not go round the earth but the earth round the sun, then we should have to proceed with great circumspection in explaining passages of Scripture which appear to teach the contrary, and rather admit that we did not understand them than declare an opinion to be false which is proved to be true.[11]

Very probably Bellarmine's liberalism is more apparent than real. The next sentence of his letter reads, "But as for myself, I shall not believe that there are such proofs until they are shown to me," and that sentence was written in full knowledge of the telescopic discoveries by which Galileo had provided strong new evidence for Copernicus' innovation. We may wonder what sort of evidence Bellarmine would have considered "real proof" against the literal word of Scripture. But he was aware, at least in principle, of the possibility of evidence that would necessitate reinterpretation. Only, by the second decade of the seventeenth century, Catholic authorities were giving greater weight to scriptural evidence and allowing less latitude for speculative dissent than they had done for centuries.

Much of the increasingly fundamentalist position that underlies the Catholic condemnation of Copernicus must, I think, be a reaction to the pressures brought to bear upon the Church by the Protestant revolt. Copernican doctrines were, in fact, condemned during the Counter Reformation, just when the Church was most convulsed by internal reforms designed to meet Protestant criticism. Anti-Copernicanism seems, at least in part, one of those reforms. Another cause of the Church's increased sensitivity to Copernicanism after 1610 may well have been a delayed awakening to the fuller theological implications of the earth's motions. In the sixteenth century those implications had rarely been made explicit. But in 1600 they were emphasized with a clamor heard throughout Europe by the execution of Giordano Bruno, the philosopher and mystic, at the stake in Rome. Bruno was not executed for Copernicanism, but for a series of theological heresies centering in his view of the Trinity, heresies for which Catholics had been executed before. He is not, as he has often been called, a martyr of science. But Bruno had found Copernicus' proposal congenial to his Neoplatonic and Democritean vision of an infinite universe containing an infinity of worlds generated by a fecund deity. He had propounded Copernicanism in England and on the Continent and had given it a significance not to be found in the *De Revolutionibus* (see Chapter 7 below). Certainly the Church feared Bruno's Copernicanism, and that fear may also have stimulated their reaction.

But whatever the reasons, the Church did, in 1616, make Copernicanism a doctrinal issue, and all the worst excesses of the battle against the earth's motion — the condemnation of Copernican opinions, the recantation and "imprisonment" of Galileo, and the dismissal and banishment of prominent Catholic Copernicans — occurred in or after that year. Once the apparatus of the Inquisition had been unleashed upon Copernicanism it was difficult to recall. Not until 1822 did the Church permit the printing of books that treated the earth's motion as physically real, and by then all but the most rigidly orthodox Protestant sects had long been persuaded. The Church's official commitment to the earth's stability did irrevocable harm both to Catholic science and, later, to Church prestige. No episode in Catholic literature has so often or so appropriately been cited against the Church as the pathetic recantation forced upon the aged Galileo in 1633.

Galileo's recantation marks the peak of the battle against Copernicanism, and, ironically, it was not delivered until a time when the outcome of the battle could be foreseen. Before 1610, when the opposition to Copernicus' doctrine was mustering, all but the most fanatical advocates of the earth's motion would have been forced to admit that the evidence for Copernicanism was weak and the counterevidence strong. Perhaps the fundamental premise of the *De Revolutionibus* would have to be abandoned. But by 1633 that was not the case. During the first decades of the seventeenth century new and stronger evidence was discovered, and the complexion of the battle changed. Even before Galileo's recantation, the new evidence had transformed the opposition to Copernicanism into a hopeless rear-guard action. The rest of this chapter examines that new evidence drawn from the heavens by three of Copernicus' immediate successors.

Tycho Brahe

If Copernicus was the greatest European astronomer in the first half of the sixteenth century, Tycho Brahe (1546–1601) was the preëminent astronomical authority of the second. And, judged purely by technical proficiency, Brahe was the greater man. But comparison is largely meaningless, because the two have different strengths and weaknesses which would not readily have merged in a single personality, and both sorts of strength were essential to the Copernican Revolution. As a cosmological and astronomical theorist, Brahe displayed a relatively traditional frame of mind. His work shows little of that Neoplatonic concern with mathematical harmonies that had been instrumental in Copernicus' break with the Ptolemaic tradition and that at the start provided the only real evidence of the earth's motion. He propounded no enduring innovations in astronomical theory. He was, in fact, a lifelong opponent of Copernicanism, and his immense prestige helped to postpone the conversion of astronomers to the new theory.

But though Brahe was no innovator of astronomical concepts, he was responsible for immense changes in the techniques of astronomical observation and in the standards of accuracy demanded from astronomical data. He was the greatest of all naked-eye observers. He designed and built many new instruments, larger, stabler, and better calibrated than those in use before. With great ingenuity he investigated and corrected many errors that developed in using these instruments, establishing a whole series of new techniques for the collection of accurate information about the position of planets and stars. Most important of all, he began the practice of making regular observations of planets as they moved through the heavens rather than observing them only when in some particularly favorable configuration. Modern telescopic observation indicates that when Brahe took particular care in determining the position of a fixed star his data were consistently accurate to 1' of arc or better, a phenomenal achievement with the naked eye. His observations of planetary position seem normally to have been reliable to about 4' of arc, more than twice the accuracy achieved by the best observers of antiquity. But even more important than the accuracy of Brahe's individual observations was the reliability and the scope of the entire body of data he collected. In his own lifetime he and the observers he trained freed European astronomy from its dependence on ancient data and eliminated a whole series of apparent astronomical problems which had derived from bad data. His observations provided a new statement of the problem of the planets, and that new statement was a prerequisite to the problem's solution. No planetary theory could have reconciled the data employed by Copernicus.

Trustworthy, extensive, and up-to-date data are Brahe's primary contribution to the solution of the problem of the planets. But he has another and a larger role in the Copernican Revolution as the author of an astronomical system that rapidly replaced the Ptolemaic system as the rallying point for those proficient astronomers who, like Brahe himself, could not accept the earth's motion. Most of Brahe's reasons for rejecting Copernicus' proposal are the usual ones, though he developed them in more detail than most of his contemporaries. But Brahe gave particular emphasis to the immense waste space that the Copernican theory opened between the sphere of Saturn and the stars merely to account for the absence of observable parallactic motion. He himself had looked for parallax with his great new instruments. Since he found none, he felt forced to reject the earth's motion. The only alternative compatible with his observations would have required a distance between the stellar sphere and Saturn seven hundred times the distance between Saturn and the sun.

But Brahe was nothing if not a proficient astronomer. Though he rejected the earth's motion, he could not ignore the mathematical harmonies which the *De Revolutionibus* had introduced into astronomy. Those new harmonies did not convert him — they were not, for him, sufficiently strong evidence to counterbalance the difficulties inherent in the earth's motion — but they must at least have increased his discontent with the Ptolemaic system, and he rejected it, too, in favor of a third system of his own invention. Brahe's system, the "Tychonic," is shown in Figure 37. Once again the earth lies stationary at the geometric center of a stellar sphere whose daily rotation accounts for the diurnal circles of the stars. As in the Ptolemaic system, the sun, moon, and planets are carried westward daily with the stars by the outer sphere, and they have additional eastward orbital motions of their own. In the diagram these orbital motions are represented by circles, though in the full Tychonic system minor epicycles, eccentrics, and equants are also required. The circles of the moon and sun are centered on the earth; to this point the system is still Ptolemaic. But the centers of the five remaining planetary orbits are transferred from the center of the earth to the sun. Brahe's system is an extension, though perhaps not a conscious one, of Heraclides' system, which attributed sun-centered orbits to Mercury and Venus.

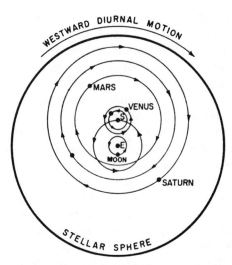

Figure 37. The Tychonic system. The earth is once again at the center of a rotating stellar sphere, and the moon and sun move in their old Ptolemaic orbits. The other planets are, however, fixed on epicycles whose common center is the sun.

The remarkable and historically significant feature of the Tychonic system is its adequacy as a compromise solution of the problems raised by the *De Revolutionibus*. Since the earth is stationary and at the center, all the main arguments against Copernicus' proposal vanish. Scripture, the laws of motion, and the absence of stellar parallax, all are reconciled by Brahe's proposal, and this reconciliation is effected without sacrificing any of Copernicus' major mathematical harmonies. The Tychonic system is, in fact, precisely equivalent mathematically to Copernicus' system. Distance determination, the apparent anomalies

in the behavior of the inferior planets, these and the other new harmonies that convinced Copernicus of the earth's motion are all preserved.

The harmonies of the Tychonic system may be developed individually and in detail by the same techniques employed in discussing
Copernicus' system, but for present purposes the following abbreviated
demonstration of the mathematical equivalence of the Copernican and
Tychonic systems should be sufficient. Imagine the sphere of the stars
in Figure 37 immensely expanded until an observer on the moving
sun could no longer observe any stellar parallax from opposite sides
of the sun's orbit. This expansion does not affect the system's mathematical account of any of the planetary motions. Now imagine that
within this expanded stellar sphere the various planets are driven
about their orbits by a clockwork mechanism like that indicated schematically in Figure 38a for the earth, the sun, and Mars. In the diagram
the sun is attached to the central earth by an arm of fixed length which
carries it counterclockwise about the earth, and Mars is attached to
the sun by another arm of fixed length which moves it clockwise about
the moving sun. Since the lengths of both arms are fixed throughout the
motion, the clockwork mechanism will produce just the circular orbits
indicated in Figure 37.

Now imagine that, without interfering with the gears that drive
the arms in Figure 38a, the whole mechanism is picked up and, with
the arms turning as before, put down again with the sun fixed at the
central position formerly held by the earth. This is the situation indicated in Figure 38b. The arms have the same lengths as before; they
are driven at the same rates by the same mechanism; and they therefore retain the same *relative* positions at each instant of time. All of
the geometric spatial relations of the earth, sun, and Mars in the diagram of Figure 38a are preserved by the arrangement of Figure 38b,
and since only the fixed point of the mechanism has been changed,
all the relative motions must be identical.

But the motions produced by the mechanism of Figure 38b are
Copernican motions. That is, the fixed arms shown in the second diagram move both the earth and Mars in circular orbits about the sun,
and those orbits are just the basic ones described by Copernicus.
Carrying out the same argument with the hypothetical mechanism
of Figure 38 elaborated to include all the planets, demonstrates that

the equivalence is general. Omitting minor epicycles and eccentrics, which have no bearing on the harmonies of Copernicus' system, the Tychonic system is transformed to the Copernican system simply by holding the sun fixed instead of the earth. The relative motions of the planets are the same in both systems, and the harmonies are therefore preserved. Mathematically the only possible difference between the motions in the two systems is a parallactic motion of the stars, and that motion was eliminated at the start by expanding the stellar sphere until parallax was imperceptible.

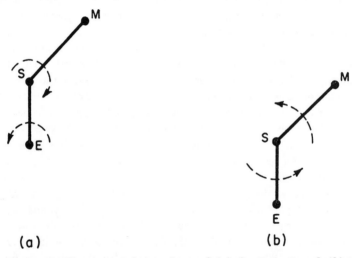

(a)   (b)

Figure 38. The geometrical equivalence of (a) the Tychonic and (b) the Copernican systems. In (a) the sun S is carried eastward about the stationary earth E by the the rigid arm ES. Simultaneously, the planet Mars, M, is carried westward about S by the steady rotation of the arm SM. Since ES rotates more rapidly than SM, the net motion of Mars is eastward except during the brief period when SM crosses over ES. In the second diagram (b) the same arms are shown rotating about the fixed sun S. The *relative positions* of E, S, and M are the same as those in (a), and they will stay the same while the arms in the two diagrams rotate. Notice particularly that in (b) the angle ESM must decrease as it does in (a) because ES rotates about the sun more rapidly than SM.

The Tychonic system has incongruities all its own: most of the planets are badly off center; the geometric center of the universe is no longer the center for most of the celestial motions; and it is difficult to imagine any physical mechanism that could produce planetary motions even approximately like Brahe's. Therefore the Tychonic system did not convert those few Neoplatonic astronomers, like Kepler, who had been attracted to Copernicus' system by its great symmetry. But it did convert most technically proficient non-Copernican astronomers of the day, because it provided an escape from a widely felt dilemma: it retained the mathematical advantages of Copernicus' system without the physical, cosmological, and theological drawbacks. That is the real importance of the Tychonic system. It was an almost perfect compromise, and in retrospect the system seems to owe its existence to the felt need for such a compromise. The Tychonic system, to which almost all the more erudite seventeenth-century Ptolemaic astronomers retreated, appears to be an immediate by-product of the *De Revolutionibus*.

Brahe himself would have denied this. He proclaimed that he had taken nothing in his system from Copernicus. But he can scarcely have been conscious of the pressures at work on him and his contemporaries. Certainly he knew both Ptolemaic and Copernican astronomy thoroughly before he thought of his own system, and he was clearly aware in advance of the predicament that his own system was to resolve. The immediate success of the system is one index of the strength and prevalence of the need. That two other astronomers disputed Brahe's priority and claimed to have worked out similar compromise solutions for themselves provides additional evidence for the role of the *De Revolutionibus* and the resulting climate of astronomical opinion in the genesis of the Tychonic system. Brahe and his system provide the first illustration of one of the major generalizations that closed the last chapter: the *De Revolutionibus* changed the state of astronomy by posing new problems for all astronomers.

Brahe's criticisms of Copernicus and his compromise solution of the problem of the planets show that, like most astronomers of his day, he was unable to break with traditional patterns of thought about the earth's motion. Among Copernicus' successors Brahe is one of the immense body of conservatives. But the effect of his work was not conservative. On the contrary, both his system and his observations forced his successors to repudiate important aspects of the Aristotelian-Ptolemaic universe and thus drove them gradually toward the Copernican camp. In the first place, Brahe's system helped to familiarize astronomers with the mathematical problems of Copernican astronomy, for geometrically the Tychonic and Copernican systems were identical. More important, Brahe's system, abetted by his observations of comets, to be discussed below, forced his followers to abandon the crystalline spheres which, in the past, had carried the planets about their orbits. In the Tychonic system, as indicated by Figure 37, the orbit of Mars intersects the orbit of the sun. Both Mars and the sun cannot, therefore, be embedded in spheres that carry them about, for the two spheres would have to penetrate and move through each other at all times. Similarly, the sun's sphere passes through the spheres of Mercury and Venus. Abandoning the crystalline spheres does not make a man a Copernican; Copernicus himself had utilized spheres to account for the planetary motions. But the spheres had, in one of a number of modifications, been an essential ingredient of the Aristotelian cosmological tradition which was the principal barrier to the success of Copernicanism. Any break with the tradition worked for the Copernicans, and the Tychonic system, for all its traditional elements, was an important break.

Brahe's skillful observations were even more important than his system in leading his contemporaries toward a new cosmology. They provided the essential basis for the work of Kepler, who converted Copernicus' innovation into the first really adequate solution of the problem of the planets. And even before they were used to revise Copernicus' system, the new data collected by Brahe suggested the necessity of another major departure from classical cosmology — they raised questions about the immutability of the heavens. Late in 1572, when Brahe was at the beginning of his career in astronomy, a new celestial body appeared in the constellation Cassiopeia, directly across the pole from the Big Dipper. When first observed it was very brilliant, as clear as Venus at its greatest brightness; during the next eighteen months the new occupant of the heavens grew gradually dimmer; and finally it vanished altogether early in 1574. From the start the new visitor drew the interest of scientists and nonscientists throughout Europe. It could not be a comet, the only sort of celestial apparition widely recognized by astronomers and astrologers, for it had no tail, and it always appeared in the same position against the sphere of the stars. Clearly it was a portent; astrological activity multiplied; and astronomers everywhere devoted their observations and their writings to the "new star" in the heavens.

The word "star" is the key to the astronomical and cosmological significance of the new phenomenon. If it were a star, then the immutable heavens had changed, and the basic contrast between the superlunary region and the corruptible earth was in question. If it were a star, the earth might more easily be conceived as a planet, for the transitory character of terrestrial affairs would now have been discovered in the heavens as well. Brahe and the best of his contemporaries did conclude that the visitor was a star. Observations like the one illustrated in Figure 39 indicated that it could not be located

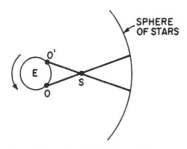

Figure 39. Diurnal parallax of a body below the stars. If S is between the earth and the sphere of the stars, then it should appear at different positions against the background of stars when observed by terrestrial observers at O and O'. Two observers are not required. The eastward rotation of the earth (or the equivalent westward rotation of the observed body and the stellar sphere) carries an observer from O to O' in six hours; as a result of the rotation the body S appears to change its position continually, returning to its starting point among the stars after twenty-four hours. If S were as close as the moon, its apparent displacement during six hours would be very nearly 1°. Bodies farther from the earth show less displacement.

With modern instruments the technique illustrated above is useful in determining the distances to the moon and planets, but naked-eye observations are not accurate enough for this application. The large size of the moon and its rapid orbital motion disguise the parallactic effect. The planets are too far away.

below the sphere of the moon or even close to the sublunary region. Probably it was among the stars, for it was observed to move with them. Another cause for cosmological upheaval had been discovered.

The sixteenth-century discovery of the mutability of the heavens might have been relatively ineffectual if the only evidence of superlunary change had been drawn from the new star, or nova, of 1572. It was a transient phenomenon; those who chose to reject Brahe's data could not be refuted; by the time the data were published the star had disappeared; and some less careful observers could always be discovered who had observed a parallax sufficient to place the nova below the moon. But fortunately additional and continuing evidence of superlunary change was provided by comets which Brahe observed carefully in 1577, 1580, 1585, 1590, 1593, and 1596. Once again no measurable parallax was observed, and the comets too were therefore located beyond the moon's sphere where they moved through the region formerly filled by the crystalline spheres.

Like the observations of the nova, Brahe's discussions of comets failed to convince all of his contemporaries. During the first decades of the seventeenth century Brahe was frequently attacked, occasionally with the same bitterness displayed toward Copernicus, by those who believed that other data proved comets and novas to be sublunary phenomena and that the inviolability of the heavens could therefore be preserved. But Brahe did convince a large number of astronomers of a basic flaw in the Aristotelian world view, and, more important, he provided a mode of argument by which skeptics could continuously check his conclusions. Comets bright enough to be seen with the naked eye appear every few years. After their superlunary character had been deduced from observation and then widely debated, the evidence that comets provided for the mutability of the heavens could not indefinitely be ignored or distorted. Once again the Copernicans were the gainers.

Somehow, in the century after Copernicus' death, all novelties of astronomical observation and theory, whether or not provided by Copernicans, turned themselves into evidence for the Copernican theory. That theory, we should say, was proving its fruitfulness. But, at least in the case of comets and novas, the proof is very strange, for the observations of comets and novas have nothing whatsoever to do with the earth's motion. They could have been made and interpreted by a Ptolemaic astronomer just as readily as by a Copernican. They are not, in any direct sense, by-products of the *De Revolutionibus*, as the Tychonic system was.

But neither can they be quite independent of the *De Revolutionibus* or at least of the climate of opinion within which it was created. Comets had been seen frequently before the last decades of the sixteenth century. New stars, though they appear less frequently to the naked eye than comets, must also have been occasionally accessible to observers before Brahe's time; one more appeared in the year before his death and a third in 1604. Even Brahe's fine instruments were not required to discover the superlunary character of novas and comets; a parallactic shift of 1° could have been measured without those instruments, and a number of Brahe's contemporaries did independently conclude that comets were superlunary using instruments that had been known for centuries. The Copernican Maestlin needed only a piece of thread to decide that the nova of 1572 was beyond the moon. In short, the observations with which Brahe and his contemporaries speeded the downfall of traditional cosmology and the rise of Copernicanism could have been made at any time since remote antiquity. The phenomena and the requisite instruments had been available for two millenniums before Brahe's birth, but the observations were not made or, if made, were not widely interpreted. During the last half of the sixteenth century age-old phenomena rapidly changed their meaning and significance. Those changes seem incomprehensible without reference to the new climate of scientific thought, one of whose first outstanding representatives is Copernicus. As suggested at the end of the last chapter, the *De Revolutionibus* marked a turning point, and there was to be no turning back.

Johannes Kepler

Brahe's work indicates that after 1543 even the opponents of Copernicanism, at least the ablest and most honest ones, could scarcely help promoting major reforms in astronomy and cosmology. Whether or not they agreed with Copernicus, he had changed their field. But the work of an anti-Copernican like Brahe does not show the extent of those changes. A better index of the novel problems that accrued to astronomy after Copernicus' death is provided by the research of Brahe's most famous colleague, Johannes Kepler (1571–1630). Kepler was a lifelong Copernican. He seems first to have been converted to the system by Maestlin when he was a student at the Protestant university of Tübingen, and his faith in it never wavered after his student days. Throughout his life he referred in the rhapsodic tones characteristic of Renaissance Neoplatonism to the suitability of the role that Copernicus had attributed to the sun. His first important book, the *Cosmographical Mystery*, published in 1596, opened with a lengthy defense of the Copernican system, emphasizing all those arguments from harmony that we discussed in Chapter 5 and adding many new ones besides: Copernicus' proposal explains why

Mars's epicycle had been so much larger than Jupiter's and Jupiter's than Saturn's; sun-centered astronomy shows why, of all the celestial wanderers, only the sun and moon fail to retrogress; and so on and on. Kepler's arguments are the same as Copernicus', though more numerous, but Kepler, in contrast to Copernicus, develops the arguments at length and with detailed diagrams. For the first time the full force of the mathematical arguments for the new astronomy was demonstrated.

But though Kepler was full of praise for the conception of a sun-centered planetary system, he was quite critical of the particular mathematical system that Copernicus had developed. Again and again Kepler's writings emphasized that Copernicus had never recognized his own riches and that after the first bold step, the transposition of the sun and earth, he had stayed too close to Ptolemy in developing the details of his system. Kepler was acutely and uncomfortably aware of the incongruous archaic residues in the *De Revolutionibus*, and he took it upon himself to eliminate them by exploiting fully the earth's new status as a planet governed, like the other planets, by the sun.

Copernicus had not quite succeeded in treating the earth as just another planet in a sun-centered system. Unlike the qualitative sketch in the First Book of the *De Revolutionibus*, the mathematical account of the planetary system developed in the later books attributed several special functions to the earth. For example, in the Ptolemaic system the planes of all planetary orbits had been constructed so that they intersected at the center of the earth, and Copernicus preserved this terrestrial function in a new form by drawing all orbital planes so that they intersected at the center of the earth's orbit. Kepler insisted that, since the sun governed the planets and the earth had no unique status, the planes of the orbits must intersect in the sun. By redesigning the Copernican system accordingly he made the first significant progress since Ptolemy in accounting for the north and south deviations of the planets from the ecliptic. Kepler had improved Copernicus' mathematical system by applying strict Copernicanism to it.

A similar insistence upon the parity of the planets enabled Kepler to eliminate a number of pseudo problems that had distorted Copernicus' work. Copernicus had, for example, believed that the eccentricities of Mercury and Venus were slowly changing, and he had added circles to his system to account for the variation. Kepler showed that the apparent change was due only to an inconsistency in Copernicus' definition of eccentricity. In the *De Revolutionibus* the eccentricity of the earth's orbit was measured from the sun (it is the distance SO_E in Figure 34a, p. 169) while the eccentricities of all other orbits were measured from the center of the earth's orbit (Mars's eccentricity is $O_E O_M$ in Figure 34b). Kepler insisted that all planetary eccentricities must, in a Copernican universe, be computed in the same way and from the sun. When the new method was incorporated in his system, several of the apparent variations of eccentricity vanished, and the number of circles required in computation was reduced.

Each of these examples shows Kepler striving to adapt Copernicus' overly Ptolemaic mathematical techniques to the Copernican vision of a sun-dominated universe, and it was by continuing this effort that Kepler finally resolved the problem of the planets, transforming Copernicus' cumbersome system into a supremely simple and accurate technique for computing planetary position. His most essential discoveries were made while studying the motion of Mars, a planet whose eccentric orbit and proximity to the earth produce irregularities that had always challenged the ingenuity of mathematical astronomers. Ptolemy had been unable to account for its motion as satisfactorily as for that of the other planets, and Copernicus had not improved on Ptolemy. Brahe had attempted a new solution, undertaking a long series of observations specially for the purpose, but surrendering the problem as he encountered its full difficulties. Kepler, who had worked with Brahe during the last years of Brahe's life, inherited the new observations and, in the years after Brahe's death, took up the problem himself.

It was an immense labor which occupied much of Kepler's time for almost ten years. Two orbits had to be worked out: the orbit of Mars itself and the orbit of the earth from which Mars is observed. Again and again Kepler was forced to change the combination of circles used in computing these orbits. System after system was tried and rejected because it failed to conform to Brahe's brilliant observations. All of the intermediate solutions were better than the systems of Ptolemy and of Copernicus; some gave errors no larger than 8' of arc, well within the limits of ancient observation. Most of the systems that Kepler discarded would have satisfied all earlier mathematical astronomers. But they had lived before Brahe, whose data were accurate to 4' of arc. To us, Kepler said, Divine goodness has given a most diligent observer in Tycho Brahe, and it is therefore right that we should with a grateful mind make use of this gift to find the true celestial motions.

A long series of unsuccessful trials forced Kepler to conclude that no system based upon compounded circles would solve the problem. Some other geometric figure must, he thought, contain the key. He tried various sorts of ovals, but none eliminated the discrepancies between his tentative theory and observation. Then, by chance, he noticed that the discrepancies themselves varied in a familiar mathematical fashion, and investigating this regularity he discovered that theory and observation could be reconciled if the planets moved in elliptical orbits with variable speeds governed by a simple law which he also specified. These are the results that Kepler announced in *On the Motion of Mars*, first published at Prague in 1609. A mathematical technique simpler than any employed since Apollonius and Hipparchus yielded predictions far more accurate than any that had ever been made before. The problem of the planets had at last been solved, and it was solved in a Copernican universe.

The two laws that constitute Kepler's (and our) final solution of the problem of the planets are described in detail in Figure 40. The planets move in simple elliptical paths, and the sun occupies one of the two foci of each elliptical orbit — that is Kepler's First Law. His Second Law follows immediately, completing the description embodied in the First — the orbital speed of each planet varies in such a way that a line joining the planet to the sun sweeps through equal areas of the ellipse in equal intervals of time. When ellipses are substituted for the basic circular orbits common to Ptolemy's and Copernicus' astronomy and when the law of equal areas is substituted for the law of uniform motion about a point at or near the center, all need for eccentrics, epicycles, equants, and other *ad hoc* devices vanishes. For the first time a single uncompounded geometric curve and a single speed law are sufficient for predictions of planetary position, and for the first time the predictions are as accurate as the observations.

The Copernican astronomical system inherited by modern science is, therefore, a joint product of Kepler and Copernicus. Kepler's system of six ellipses made sun-centered astronomy work, displaying simultaneously the economy and the fruitfulness implicit in Copernicus' innovation. We must try to discover what was required for this transition of the Copernican system to its modern, Keplerian, form. Two of the prerequisites of Kepler's work are already apparent. He had to be a convinced Copernican, a man who would begin his search for more adequate orbits by treating the earth as a mere planet and who would construct the planes of all planetary orbits through the center of the sun. In addition, he needed Brahe's data. The data used by Copernicus and his European predecessors were too infected with errors to be

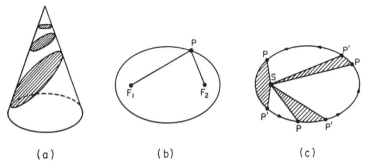

(a) (b) (c)

Figure 40. Kepler's first two Laws. Diagrams (a) and (b) define the ellipse, the geometric curve in which all planets that obey Kepler's First Law must move. In (a) the ellipse is shown as the closed curve in which a plane intersects a circular cone. When the plane is perpendicular to the axis of the cone, the intersection is a circle, a special case of the ellipse. As the plane is tilted, the curve of intersection is elongated into more typically elliptical patterns.

A more modern and somewhat more useful definition of the ellipse is given in diagram (b). If two ends of a slack string are attached to two points F_1 and F_2 in a plane, and if a pencil P is inserted into the slack and then moved so that it just keeps the string taut at all times, the point of the pencil will generate an ellipse. Changing the length of the string or moving the foci F_1 and F_2 together or apart alters the shape of the ellipse in the same way as a change in the tilt of the plane in diagram (a). Most planetary orbits are very nearly circular, and the foci of the corresponding ellipses are therefore quite close together.

Diagram (c) illustrates Kepler's Second Law, which governs orbital speed. The sun is at one focus of the ellipse, as required by the First Law, and its center is joined by straight lines to a number of planetary positions P and P', arranged so that each of the three shaded sectors SPP' has the same area. The Second Law states that, since each of these areas is the same, the planet must move through each of the corresponding arcs PP' in equal times. When near the sun, the planet must move relatively quickly so that the short line SP will sweep out the same area per unit time as is swept out by the longer line SP when the planet is moving more slowly farther from the sun.

explained by any set of simple orbits, and even if freed from error they would not have sufficed. Observations less precise than Brahe's could have been explained, as Kepler himself showed, by a classical system of compounded circles. The process by which Kepler arrived at his famous Laws depends, however, upon more than the availability of accurate data and a prior commitment to the planetary earth. Kepler was an ardent Neoplatonist. He believed that mathematically simple laws are the basis of all natural phenomena and that the sun is the physical cause of all celestial motions. Both his most lasting and his most evanescent contributions to astronomy display these two aspects of his frequently mystical Neoplatonic faith.

In a passage quoted at the end of Chapter 4 Kepler described the sun as the body "who alone appears, by virtue of his dignity and power, suited . . . [to move the planets in their orbits], and worthy to become the home of God himself, not to say the first mover." This conviction, together with certain intrinsic incongruities discussed above, was his reason for rejecting the Tychonic system. It also played an immensely important role in his own research, particularly in his derivation of the Second Law upon which the First depends. In its origin the Second Law is independent of any but the crudest sort of observation. It arises rather from Kepler's physical intuition that the planets are pushed around their orbits by rays of a moving force, the *anima motrix*, which emanates from the sun. These rays must, Kepler believed, be restricted to the plane of the ecliptic, in or near which all the planets moved. Therefore the number of rays that impinged on a planet and the corresponding force that drove the planet around the sun would decrease as the distance between the planet and the sun increased. At twice the distance from the sun half as many rays of the *anima motrix* would fall on a planet (Figure 41*a*), and the velocity of the planet in its orbit would, in consequence, be half of its orbital velocity at its original distance from the sun. A planet, *P*, moving about the sun, *S*, on an eccentric circle (Figure 41*b*) or some other closed curve must move at a speed inversely proportional to *SP*. The speed will be greatest when the planet is at the perihelion, *p*, closest to the sun, and least at the aphelion, *a*, where the planet is farthest from the sun. As the planet moves around the orbit, its speed will vary continually between these extremes.

Long before he began to work on elliptical orbits or stated the law of areas in its familiar modern form, Kepler had worked out this inverse-distance speed law to replace both the ancient law of uniform circular motion and the Ptolemaic variant which permitted uniform motion with respect to an equant point. This early speed law was very much "pulled from a hat" by a strange intuition — one that was rapidly discarded by his successors — of the forces that must govern a sun-dominated universe. Furthermore, this early form is not quite correct. The later law of areas, Kepler's so-called Second Law, is not quite equivalent to the inverse-distance law, and the law of areas gives somewhat better results. But when used to compute planetary position the two forms of the speed law lead to almost the same predictions. Kepler mistakenly thought the two equivalent in principle and used them interchangeably throughout his life. For all its visionary over-tones the early Neoplatonic speed law proved fundamental in Kepler's most fruitful research.

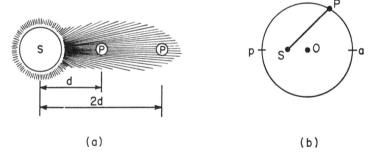

(a)   (b)

Figure 41. Kepler's earliest speed law. Diagram (a), which shows typical rays of the *anima motrix* radiating from the sun, illustrates the physical theory from which Kepler derived the law. Diagram (b) shows how the law could be applied to a planet moving on an eccentric circle.

Unlike his derivation of the speed law, Kepler's work on elliptical orbits was completely dependent upon the most painstaking and exhaustive study of the best available astronomical observations. Trial orbit after trial orbit had to be abandoned because, after laborious computation, it did not quite match Brahe's data. Kepler's scrupulous attempt to fit his orbits to objective data is often cited as an early example of the scientific method at its best. Yet even the law of elliptical orbits, Kepler's First Law, was not derived from observation and computation alone. Unless the planetary orbits are assumed to be precisely reëntrant (as they were after Kepler's work but not before), a speed law is required to compute orbital shape from naked-eye data. When analyzing Brahe's observations, Kepler made constant use of his earlier Neoplatonic guess.

The interrelation of orbit, speed law, and observation was obscured in our earlier discussions of astronomical theory, because ancient and medieval astronomers chose a simple speed law in advance. Before Kepler astronomers assumed that each of the compounded circles which moved a planet around its orbit must rotate uniformly with respect to a point at or near its center. Without some such assumption they could not have begun the elaboration of orbits to fit observations, for in the absence of a speed law the specification of an orbit tells little or nothing about where a planet will appear among the stars at a particular time. Neither speed law nor orbit can be independently derived from or checked against observation. Therefore, when Kepler rejected the ancient law of uniform motion, he had to replace it or else abandon planetary computations entirely. In fact, he rejected the ancient law only after (and probably because) he had developed a law of his own — a law that his Neoplatonic intuition told him was better suited than its ancient counterpart to govern celestial motions in a sun-dominated universe.

Kepler's derivation of the inverse-distance law displays his belief in mathematical harmonies as well as his faith in the causal role of the sun. Having developed the conception of the *anima motrix* Kepler insisted that it must operate in the simplest way compatible with crude observation. He knew, for example, that planets move fastest at perihelion, but he had few other data, none of them quantitative, on which to base an inverse-distance law. But Kepler's belief in number harmonies and the role of this belief in his work is more forcefully exhibited in another one of the laws that modern astronomy inherits from him. This is Kepler's so-called Third Law, announced during 1619 in the *Harmonies of the World.*

The Third Law was a new sort of astronomical law. Like their ancient and medieval counterparts the First and Second Laws govern only the motions of individual planets in their individual orbits. The Third Law, in contrast, established a relation between the speeds of planets in different orbits. It states that if T_1 and T_2 are the periods that two planets require to complete their respective orbits once, and if R_1 and R_2 are the average distances between the corresponding planets and the sun, then the ratio of the squares of the orbital periods is equal to the ratio of the cubes of the average distances from the sun, or $(T_1 / T_2)^2 = (R_1 / R_2)^3$. This is a fascinating law, for it points to a regularity never before perceived in the planetary system. But, at least in Kepler's day, that was all it did. The Third Law did not, in itself, change the theory of the planets, and it did not permit astronomers to compute any quantities that were previously unknown. The sizes and the periods associated with each planetary orbit were available in advance.

But though it had little immediate practical use, the Third Law is just the sort of law that most fascinated Kepler throughout his career. He was a mathematical Neoplatonist or Neopythagorean who believed that all of nature exemplified simple mathematical regularities which it was the scientists' task to discover. To Kepler and others of his turn of mind a simple mathematical regularity was itself an explanation. To him the Third Law in and of itself explained why the planetary orbits had been laid out by God in the particular way that they had, and that sort of explanation, derived from mathematical harmony, is what Kepler continually sought in the heavens. He propounded a number of other laws of the same kind, laws which we have since abandoned because, though harmonious, they do not fit observation well enough to seem significant. But Kepler was not so selective. He thought that he had discovered and demonstrated a large number of these mathematical regularities, and they were his favorite astronomical laws.

In Kepler's first major work, the *Cosmographical Mystery*, he argued that both the number of the planets and the size of their orbits could be understood in terms of the relation between the planetary spheres and the five regular or "cosmic" solids. These are the solids shown in Figure 42*a*, and they have the unique characteristic that all of the faces of each solid are identical and that only equilateral figures are used for faces. It had been shown in antiquity that there could be only five such solids: cube, tetrahedron, dodecahedron, icosahedron, and octahedron. Kepler proclaimed that if the sphere of Saturn were circumscribed about the cube within which Jupiter's sphere was inscribed, and if the tetrahedron were placed just inside Jupiter's sphere with Mars's sphere inscribed in it, and so on for the three other solids and three other spheres, then the relative dimensions of all the spheres would be just those that Copernicus had determined by measurement. The construction is shown in Figure 42*b*. If it is to be used, there can be only six planets, corresponding to the five regular solids, and when it is used the permissible relative dimensions of the

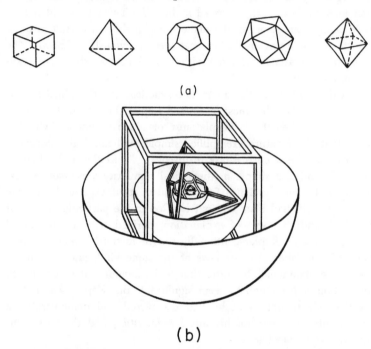

(a)

(b)

Figure 42. Kepler's application of the five regular solids. Diagram (*a*) shows the solids themselves. From left to right they are: cube, tetrahedron, dodecahedron, icosahedron, and octahedron. Their order is the one that Kepler developed to account for the sizes of the planetary spheres. Diagram (*b*) shows the solids in this application. Saturn's sphere is circumscribed about the cube, and Jupiter's sphere is inscribed in it. The tetrahedron is inscribed in Jupiter's sphere, and so on.

planetary spheres are determined. That, said Kepler, is why there are only six planets and why they are arranged as they are. God's nature is mathematical.

Kepler's use of the regular solids was not simply a youthful extravagance, or if it was, he never grew up. A modified form of the same law appeared twenty years later in his *Harmonies of the World*, the same book that propounded the Third Law. Also in that book Kepler elaborated a new set of Neoplatonic regularities which related the maximum and minimum orbital speeds of the planets to the concordant intervals of the musical scale. Today this intense faith in number harmonies seems strange, but that is at least partly because today scientists are prepared to find their harmonies more abstruse. Kepler's application of the faith in harmonies may seem naïve, but the faith itself is not essentially different from that motivating bits of the best contemporary research. Certainly the scientific attitude demonstrated in those of Kepler's "laws" which we have now discarded is not distinguishable from the attitude which drove him to the three Laws that we now retain. Both sets, the "laws" and the Laws, arise from the same renewed faith in the existence of mathematical harmony that had so large a role in driving Copernicus to break with the astronomical tradition and in persuading him that the earth was, indeed, in motion. But in Kepler's work, and particularly in the parts of it that we have now discarded, the Neoplatonic drive to discover the hidden mathematical harmonies embedded in nature by the Divine Spirit are illustrated in a purer and more distinct form.

Galileo Galilei

Kepler solved the problem of the planets. Ultimately his version of Copernicus' proposal would almost certainly have converted all astronomers to Copernicanism, particularly after 1627 when Kepler issued the *Rudolphine Tables*, derived from his new theory and clearly superior to all the astronomical tables in use before. The story of the astronomical components of the Copernican Revolution might therefore end with the gradual acceptance of Kepler's work because that work contains all the elements required to make the Revolution in astronomy endure. But, in fact, the astronomical components of the story do not end there. In 1609 the Italian scientist Galileo Galilei (1564–1642) viewed the heavens through a telescope for the first time, and as a result contributed to astronomy the first qualitatively new sort of data that it had acquired since antiquity. Galileo's telescope changed the terms of the riddle that the heavens presented to astronomers, and it made the riddle vastly easier to solve, for in Galileo's hands the telescope disclosed countless evidences for Copernicanism. But Galileo's new statement of the riddle was not formulated until after the riddle had been solved by other means. If it had been announced earlier, the story of the Copernican Revolution would be quite different. Coming when it did, Galileo's astronomical work contributed primarily to a mopping-up operation, conducted after the victory was clearly in sight.

In 1609 the telescope was a new instrument, though it is not clear just how new it was. Galileo heard that some Dutch lens grinder had combined two lenses in a way that magnified distant objects; he tried various combinations himself and quickly produced a low-power telescope of his own. Then he did something which, apparently, no one had done before: he directed his glass to the heavens, and the result was astounding. Every observation disclosed new and unsuspected objects in the sky. Even when the telescope was directed to familiar celestial objects, the sun, moon, and planets, remarkable new aspects of these old friends were discovered. Galileo, who had been a Copernican for some years before he knew of the telescope, managed to turn each new discovery into an argument for Copernicanism.

The telescope's first disclosure was the new worlds in the firmament about which Donne, only two years later, complained. Wherever he turned his glass, Galileo found new stars. The population of the most crowded constellations increased. The Milky Way, which to the naked eye is just a pale glow in the sky (it had frequently been explained as a sublunary phenomenon, like comets, or as a reflection of diffused light from the sun and moon) was now discovered to be a gigantic collection of stars, too dim and too little separated to be resolved by the naked eye. Overnight the heavens were crowded by countless new residents. The vast expansion of the universe, perhaps its infinitude, postulated by some of the Copernicans, seemed suddenly less unreasonable. Bruno's mystical vision of a universe whose infinite extent and population proclaimed the infinite procreativeness of the Deity was very nearly transformed into a sense datum.

Observation of the stars also resolved a more technical difficulty that had confronted the Copernicans. Naked-eye observers had estimated the angular diameter of stars and, with the aid of the accepted figure for the distance between the earth and the sphere of the stars, had transformed the angular diameter into an estimate of linear dimensions. In a Ptolemaic universe these estimates had given not unreasonable results: the stars might be as large as the sun, or thereabouts. But, as Brahe repeatedly emphasized in his attacks upon Copernicanism, if the Copernican universe were as large as the absence of stellar parallax demanded, then the stars must be incredibly large. The brighter stars of the heavens must, Brahe computed, be so large that they would more than fill the entire orbit of the earth, and this he not unnaturally refused to believe. But when the telescope was directed to the heavens, Brahe's problem turned out to be an apparent problem only. The stars did not need to be so large as he had estimated. Though the telescope immensely increased the number of stars visible in the skies, it did not increase their apparent size. Unlike the sun, moon, and planets, all of which were magnified by Galileo's glass, the stars retained the size they had had before. It became apparent that the angular diameter of stars had been immensely overestimated by naked-eye observation, an error now explained as a consequence of atmospheric turbulence which blurs the images of stars and spreads them over a wider area in the eye than would be covered by their undistorted image alone. The same effect makes the stars seem to twinkle; it is largely suppressed by the telescope, which gathers a larger number of rays to the eye.

The stars did not, however, provide the only, or even the best, evidence for Copernicanism. When Galileo turned his telescope to the moon, he found that its surface was covered by pits and craters, valleys and mountains. Measuring the length of the shadows cast into craters and by mountains at a time when the relative positions of the sun, moon, and earth were known, he was able to estimate the depths of the moon's declivities and the height of its protuberances and to begin a three-dimensional description of the moon's topography. It was not, Galileo decided, very different from the earth's topography. Therefore, like the measurements of the parallax of comets, telescopic observations of the moon raised doubts about the traditional distinction between the terrestrial and the celestial regions, and those doubts were reinforced almost immediately by telescopic observations of the sun. It too showed imperfections, dark spots which appeared and disappeared on its surface. The very existence of the spots conflicted with the perfection of the celestial region; their appearance and disappearance conflicted with the immutability of the heavens; and, worst of all, the motion of the spots across the sun's disk indicated that the sun rotated continually on its axis and thus provided a visible paradigm for the axial rotation of the earth.

But this was not the worst. Galileo looked at Jupiter with his telescope and discovered four small points of light quite close to it in the sky. Observations made on successive nights showed that they continually rearranged their relative positions in a manner that could most simply be explained by supposing that they revolved continually and quite rapidly about Jupiter (Figure 43). These bodies were

Figure 43. Three successive observations of Jupiter and its satellites separated by intervals of several days. The constant rearrangement of the four small satellites is most easily explained by supposing that the satellites are constantly rotating about the larger planet.

the four principal moons of Jupiter, and their discovery had an immense impact upon the seventeenth-century imagination. There were, it appeared, new worlds "in the Planets" as well as in "the Firmament." More important, these new worlds could not be conceived, on either the Ptolemaic or the Copernican hypothesis, to move in roughly circular orbits about the center of the universe. Apparently they moved around a planet, and their behavior was therefore the same as that of the earth's moon in Copernican astronomy. The discovery of Jupiter's moons therefore reduced the force of one more objection to the Copernican system. The old astronomy, as well as the new, would have to admit the existence of satellites, governed by planets. In addition, and perhaps most consequential of all, the observations of Jupiter provided a visible model of the Copernican solar system itself. Here in planetary space was a heavenly body surrounded by its own "planets," just as the planets previously known encircled the sun. The arguments for Copernicanism were multiplied by the telescope almost as rapidly as the heavenly bodies themselves.

Many other arguments were derived from telescopic observation, but only the observations of Venus provide sufficiently direct evidence for Copernicus' proposal to concern us here. Copernicus himself had noted in Chapter 10 of the First Book of the *De Revolutionibus* that the appearance of Venus could, if observable in detail, provide direct information about the shape of Venus's orbit. If Venus is attached to an epicycle moving on an earth-centered deferent, and if the center of the epicycle is always aligned with the sun, then, as indicated by Figure 44a, an observer on the earth should never be able to see more than a crescent edge of the planet. But if Venus's orbit encircles the sun as in Figure 44b, then an earthbound observer should be able to see an almost complete cycle of phases, like the moon's; only phases near "new" and "full" would be imperceptible, because Venus would then be too close to the sun. Venus's phases can not be distinguished with the naked eye, which sees the planets as mere shapeless points. But the telescope enlarges planets sufficiently to give them shape,

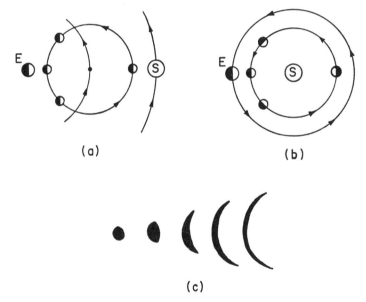

Figure 44. The phases of Venus in (a) the Ptolemaic system, (b) the Copernican system, and (c) as observed with a low-power telescope. In (a) an observer on the earth should never see more than a thin crescent of the lighted face. In (b) he should see almost the whole face of Venus illuminated just before or after Venus crosses behind the sun. This almost circular silhouette of Venus when it first becomes visible as an evening star is drawn from observations with a low-power telescope on the left of diagram (c). The successive observations drawn on the right show how Venus wanes and simultaneously increases in size as its orbital motion brings it closer to the earth.

and, as indicated in Figure 44c, its shape provides strong evidence that Venus moves in a sun-centered orbit.

The evidence for Copernicanism provided by Galileo's telescope is forceful, but it is also strange. None of the observations discussed above, except perhaps the last, provides direct evidence for the main tenets of Copernicus' theory — the central position of the sun or the motion of the planets about it. Either the Ptolemaic or the Tychonic universe contains enough space for the newly discovered stars; either can be modified to allow for imperfections in the heavens and for satellites attached to celestial bodies; the Tychonic system, at least, provides as good an explanation as the Copernican for the observed phases of and distance to Venus. Therefore, the telescope did not prove the validity of Copernicus' conceptual scheme. But it did provide an immensely effective weapon for the battle. It was not proof, but it was propaganda.

After 1609 the main psychological force of the Ptolemaic system was its conservatism. Those who held to it would not be forced to learn new ways. But if the Ptolemaic system required extensive revisions to adjust it to the results of telescopic observation, it would lose even its conservative appeal. It was very nearly as easy to make the full transition to Copernicanism as to adjust to the requisite new version of Ptolemy, and many of those who took the observations seriously did make the full transition. These new converts may also have been impelled by another consideration: the Copernicans, or at least the cosmologically more radical ones, had anticipated the sort of universe that the telescope was disclosing. They had predicted a detail, the phases of Venus, with precision. More important, they had anticipated, at least vaguely, the imperfections and the vastly increased population of the heavens. Their vision of the universe showed marked parallels to the universe that the telescope made manifest. There are few phrases more annoying or more effective than "I told you so."

For the astronomically initiate the evidence of the telescope was, perhaps, superfluous. Kepler's Laws and his *Rudolphine Tables* would have been equally, though far more slowly, effective. But it is not on the astronomically initiate that the telescope had the greatest immediate impact. The first unique role of the telescope was providing generally accessible and nonmathematical documentation for the

Copernican point of view. After 1609 men who knew only a smattering of astronomy could look through a telescope and see for themselves that the universe did not conform to the naïve precepts of common sense, and during the seventeenth century they did look. The telescope became a popular toy. Men who had never before shown interest in astronomy or in any science bought or borrowed the new instrument and eagerly scanned the heavens on clear nights. The amateur observer became a well-known figure, a subject for both emulation and parody. With him came a new literature. The beginnings of both popular science and science fiction are to be discovered in the seventeenth century, and at the start the telescope and its discoveries were the most prominent subjects. That is the greatest importance of Galileo's astronomical work: it popularized astronomy, and the astronomy that it popularized was Copernican.

The Decline of Ptolemaic Astronomy

Kepler's ellipses and Galileo's telescope did not immediately crush the opposition to Copernicanism. On the contrary, as we noted at the start of this chapter, the bitterest and most vociferous opposition was not organized until after both Kepler and Galileo had made their principal astronomical discoveries. Kepler's work, like Copernicus' sixty-five years earlier, was accessible only to trained astronomers, and, in spite of the great accuracy that Kepler was known to have achieved, many astronomers found his noncircular orbits and his new techniques for determining planetary velocities too strange and uncongenial for immediate acceptance. Until after the middle of the century a number of eminent European astronomers can be found trying to show that Kepler's accuracy can be duplicated with mathematically less radical systems. One tried to revert to epicycles; another consented to ellipses but insisted that the speed of a planet was uniform with respect to the unoccupied focus of the ellipse; still others tried orbits of another shape. None of these attempts was successful, and as the century continued fewer and fewer of them were made. But not until the last decades of the seventeenth century did Kepler's Laws become the universally accepted basis for planetary computations even among the best practicing European astronomers.

Galileo's observations met initially even greater opposition, though from a different group. With the advent of the telescope Copernicanism ceased to be esoteric. It was no longer primarily the concern of highly trained mathematical astronomers. Therefore it became more disquieting and, to some, more dangerous. The new worlds discovered by the telescope were a primary source of Donne's malaise. A few years later telescopic observations provided part of the impetus necessary to set in motion the ecclesiastical machinery of official Catholic opposition to Copernicanism. After Galileo had announced his observations in 1610, Copernicanism could not be dismissed as a mere mathematical device, useful but without physical import. Nor could even the most optimistic still regard the concept of the earth's motion as a temporary lunacy likely to vanish naturally if left to itself. The telescopic discoveries therefore provided a natural and appropriate focus for much of the continuing opposition to Copernicus' proposal. They showed the real cosmological issues at stake more quickly and more clearly than pages of mathematics.

The opposition took varied forms. A few of Galileo's more fanatical opponents refused even to look through the new instrument, asserting that if God had meant man to use such a contrivance in acquiring knowledge, He would have endowed men with telescopic eyes. Others looked willingly or even eagerly, acknowledged the new phenomena, but claimed that the new objects were not in the sky at all; they were apparitions caused by the telescope itself. Most of Galileo's opponents behaved more rationally. Like Bellarmine, they agreed that the phenomena were in the sky but denied that they proved Galileo's contentions. In this, of course, they were quite right. Though the telescope argued much, it proved nothing.

The continuing opposition to the results of telescopic observation is symptomatic of the deeper-seated and longer-lasting opposition to Copernicanism during the seventeenth century. Both derived from the same source, a subconscious reluctance to assent in the destruction of a cosmology that for centuries had been the basis of everyday practical and spiritual life. The conceptual reorientation that, after Kepler and Galileo, meant economy to scientists frequently meant a loss of conceptual coherence to men like Donne and Milton whose primary concerns were in other fields, and some men whose first interests were religious, moral, or aesthetic continued to oppose Copernicanism bitterly for a very long time. The attacks were scarcely abated by the middle of the seventeenth century. Many important tracts insisting on a literal interpretation of Scripture and upon the absurdity of the earth's motion continued to appear during the first decades of the eighteenth century. As late as 1873 the ex-president of an American Lutheran teachers' seminary published a work condemning Copernicus, Newton, and a distinguished series of subsequent astronomers for their divergence from scriptural cosmology. Even today the newspapers occasionally report the dicta of a dotard who insists upon the uniqueness and stability of the earth. Old conceptual schemes never die!

But old conceptual schemes do fade away, and the gradual extinction of the concept of the earth's uniqueness and stability clearly, if almost imperceptibly, dates from the work of Kepler and Galileo. During the century and a half following Galileo's death in 1642, a belief in the earth-centered universe was gradually transformed from an essential sign of sanity to an index, first, of inflexible conservatism, then of excessive parochialism, and finally of complete fanaticism. By the middle of the seventeenth century it is difficult to find an important astronomer who is not Copernican; by the end of the century it is impossible. Elementary astronomy responded more slowly, but during the closing decades of the century Copernican, Ptolemaic, and Tychonic astronomy were taught side by side in many prominent Protestant universities, and during the eighteenth century lectures on the last two systems were gradually dropped. Popular cosmology felt the impact of Copernicanism most slowly of all; most of the eighteenth century was required to endow the populace and its teachers with a new common sense and to make the Copernican universe the common property of Western man. The triumph of Copernicanism was a gradual process, and its rate varied greatly with social status, professional affiliation, and religious belief. But for all its difficulties and vagaries it was an inevitable process. At least it was as inevitable as any process known to the historian of ideas.

The Copernican universe assimilated during the century and a half after Galileo's death was not, however, the universe of Copernicus or even of Galileo and Kepler. Nor was its novel structure derived predominantly from astronomical evidence. Copernicus and the astronomers who followed him made the first successful substantive break with Aristotelian cosmology, and they began the construction of the new universe. But the early Copernicans did not fully see where their work was leading. During the seventeenth century many other scientific and cosmological currents converged to modify and complete the cosmological framework that had directed their thought. The Copernicanism that the eighteenth, nineteenth, and twentieth centuries inherited is a Copernicanism rebuilt to suit the seventeenth-century conception of a Newtonian world machine. That final historic integration of Copernican astronomy into the complete and coherent universe envisaged by the seventeenth century is the subject of our final chapter, though we shall treat it only with the limited detail and foreshortened perspective appropriate to an epilogue. In so far as the Copernican Revolution was a revolution merely in astronomical thought, its story ends here. What follows is a partial sketch of the larger revolution in science and cosmology — a revolution which began with Copernicus and through which the Copernican Revolution was at last completed.

7

THE NEW UNIVERSE

The New Scientific Perspective

Kepler and Galileo compiled impressive evidence for the earth's status as a moving planet. The concept of elliptical orbits and the new data collected with telescopes were, however, only *astronomical* evidence *for* the planetary earth. They did not answer the *non-astronomical* evidence *against* it. While they remained unanswered, each of those arguments, whether physical, or cosmological, or religious, testified to an immense disparity between the concepts of technical astronomy and those employed in other sciences and in philosophy. The more difficult it became to doubt the astronomical innovation, the more urgent was the need for adjustments in other fields of thought. Until those adjustments were made, the Copernican Revolution was incomplete.

Most large-scale upheavals in scientific thought produce similar conceptual disparities. We are today, for example, in the late stages of a scientific revolution initiated by Planck, Einstein, and Bohr. Their new concepts and others upon which the contemporary revolution depends show close historical parallels to Copernicus' concept of a planetary earth. Conceptions like Bohr's atom and Einstein's finite but unbounded space were introduced to solve pressing problems in a single scientific specialty. Those who accepted them did so initially because of the immense felt need in the field of their origin and in spite of their obvious conflict with common sense, physical intuition, and the basic concepts of other sciences. For a time they were used by the specialist even though, within the larger climate of scientific thought, they seemed incredible.

Continued use, however, makes even the strangest conception plausible, and once plausible the new conception gains a larger scientific function. It ceases, in the vocabulary of Chapter 1, to be merely a paradoxical and *ad hoc* device for economically describing the known, and becomes instead a basic tool for explaining and exploring nature. At this stage the new conception cannot be restricted to a single scientific specialty. Nature ought not display incompatible properties in different fields. If the physicist's electron can leap from path to path without crossing the intervening space, then the chemist's electron should have the same ability, and the philosopher's concept of matter and space demand reëxamination. Every fundamental innovation in a scientific specialty inevitably transforms neighboring sciences and, more slowly, the worlds of the philosopher and the educated layman.

Copernicus' innovation is no exception. In the early decades of the seventeenth century it was at best an astronomical innovation. Outside of astronomy it raised a host of problems just as perplexing and far more obvious than the questions of numerical detail it had resolved. Why do heavy bodies always fall toward the surface of a spinning earth as the earth moves in its orbit about the sun? How far away are the stars, and what is their role in the structure of the universe? What moves the planets, and how, in the absence of spheres, are they kept in their orbits? Copernican astronomy destroyed traditional answers to these questions, but it supplied no substitutes. A new physics and a new cosmology were required before astronomy could again participate plausibly in a unified pattern of thought.

Before the end of the seventeenth century that new science and cosmology had been created, and the men responsible were all members of the Copernican minority. Their adherence to Copernicanism gave a new shape and direction to much of their research. It provided a new set of problems, one of which — what moves the earth? — has already emerged briefly in our study of Kepler's *anima motrix*. In addition, Copernicanism supplied a multiplicity of hints about the concepts and techniques that the solution of these new problems demanded. By suggesting, for example, the unification of terrestrial and celestial laws, it made the projectile a legitimate source of information about planetary motions. Finally, Copernicanism gave a new meaning and value to a number of cosmological doctrines which, though current as minority views in antiquity and the Middle Ages, had previously been disregarded by most scientists. During the seventeenth century several of these newly popular views, particularly atomism, were a constant source of significant suggestions for science.

These new problems, new techniques, and new evaluations constitute the new perspective that seventeenth-century science gained from Copernicanism. The last chapter displayed the effects of this new perspective upon astronomy. This one will show its role in the development of other sciences and of cosmology, for the Newtonian universe was born in an intellectual climate that Copernicanism had helped make fertile. But unlike Kepler's Laws, which are the astronomical culmination of the Copernican Revolution, the Newtonian universe is a product of more than Copernicus' innovation. In discussing its evolution and discovering how the concept of a planetary earth came at last to make coherent sense we shall therefore have occasionally to introduce concepts and techniques that have been neglected to this point because they had little bearing upon the development of astronomy or cosmology until after Copernicus' death. Our problem ₃ow becomes larger than the Copernican Revolution proper.

Toward an Infinite Universe

The Aristotelian universe had been, in most versions, a finite universe — matter and space ended together at the sphere of the stars — and most early Copernicans preserved this traditional feature. In the cosmologies of Copernicus, Kepler, and Galileo the center of the sun coincided with the center of the finite stellar sphere; the sun simply changed places with the earth, becoming the unique central body, the Neoplatonic symbol of the Deity. This new two-sphere universe was a natural revision of traditional cosmology. Since there was no concrete counterevidence, it might well have endured until the nineteenth century when improved telescopes showed that different stars were at very different distances from the sun.

The role of the two-sphere construction was, however, very different in the Aristotelian and Copernican world views; finitude, in particular, had essential functions in the first that were completely absent in the second. For example, in Aristotelian science the stellar sphere was needed to carry the stars in their diurnal circles and to provide the push that kept both the planets and terrestrial objects in motion. In addition, the outer sphere defined an absolute center of space, the center toward which all heavy objects moved of their own accord. Copernicus' universe deprived the sphere of all these functions and of others as well. Terrestrial motion did not demand an absolute center of space; stones fell toward the moving earth. Nor was an outer sphere needed to produce celestial motions; whether or not located on a sphere, the stars were at rest. Copernicans were still at liberty to preserve the stellar sphere, but only tradition could supply a motive for doing so. The sphere could be abandoned without disrupting either Copernican physics or cosmology.

Copernicanism therefore allowed a new freedom to cosmological thought, and the result was a new speculative conception of the universe that would surely have horrified both Copernicus and Kepler. A century after Copernicus' death his two-sphere framework had been replaced by a universe in which the stars were scattered here and there through an infinite space. Each of these stars was a "sun," and many of them were thought to possess their own planetary systems. By 1700 the unique earth, which Copernicus had reduced to but one of six planets, had become little more than a speck of cosmic dust.

Though historians still know little about the way in which this new Copernican conception was established, its origin is quite clear. By doing away with the cosmological functions of the outer sphere, Copernicus revitalized three earlier speculative conceptions of the universe, conceptions associated respectively with scholasticism, Neoplatonism, and atomism. Before the *De Revolutionibus* these three cosmologies were quite different in structure and motive, and none was relevant to celestial science. But all were transformed to scientific cosmologies by Copernicanism, and once transformed they showed remarkable structural similarities.

Consider first the most prevalent pre-Copernican conception of an infinite universe, developed by Islamic philosophers who could not accept Aristotle's proof of the logical impossibility of a void. This universe was substantially the same as Aristotle's from the central earth to the rotating stellar sphere, but beyond the sphere space no longer ceased to exist with matter. Instead, the entire Aristotelian universe was embedded as a kernel at the center of an infinite space devoid of matter, the abode of God and his angels. Because it did not constrain God's power to make an infinite universe, this conception became relatively popular in Europe after the thirteenth century. It was described in several well-known elementary books current during Copernicus' lifetime, and knowledge of the conception may have helped him to justify the expansion of the stellar sphere that was necessitated by the absence of observed parallax. But before Copernicus, this version of an infinite universe had had little bearing upon the practice of astronomy or any other science. So long as celestial bodies were conceived to be in continuous motion, they could not easily be placed in the infinite space beyond the outermost sphere. The functions of that space were theological, not physical or astronomical.

By bringing the stars to rest, however, Copernicus made it possible to give infinite space astronomical functions, and that new freedom was first exploited about a generation after the publication of the *De Revolutionibus*. In 1576 the English Copernican, Thomas Digges, introduced the idea of an infinite universe into an otherwise straightforward paraphrase of Copernicus' Book One, and the result, reproduced from Digges's original illustration, is shown in Figure 45. The central core of the universe is identical with the universe of the *De Revolutionibus*, but the stars have been removed from the surface of the stationary stellar sphere and scattered outward through the infinite space posited by the older minority cosmological tradition. Though few of Copernicus' immediate successors went as far as Digges, many of them recognized that the stars no longer had to be on a sphere and that the distances between individual stars and the sun might vary without affecting the appearances. When Galileo's telescope showed countless new stars where none had been seen before, the scattering of stars through an immeasurably distant space seemed to the less traditionally minded astronomers almost a matter of experience.

Digges was the first to describe an infinite Copernican universe, but he achieved infinity only by the unconscious introduction of a paradox which, in antiquity and the Middle Ages, had provided a principal reason for rejecting infinite space. Digges's unique central sun is a contradiction, for it is no more "at the center" than each one of the stars and planets. The center is the point that is equidistant from all points on the periphery, and that condition is satisfied by every point in an infinite universe or by none. This paradox had been fully elaborated a century before Copernicus by the important Neoplatonist, Nicholas of Cusa. He had believed that the universe was an infinite sphere — no smaller sphere would, he said, be consistent with God's creative omnipotence — and he had expressed the resulting paradox by declaring that the center of the sphere everywhere coincided with its periphery. In his universe each body, fixed or moving, was simultaneously at the center, on the surface, and in the interior. Be-

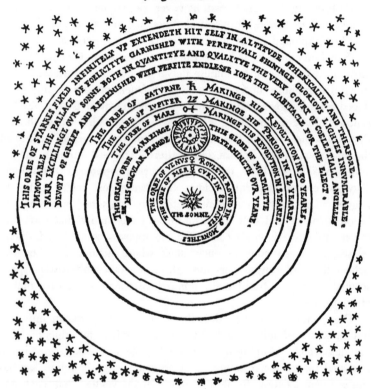

Figure 45. The infinite Copernican universe of Thomas Digges, reproduced from his *Perfit Description of the Caelestiall Orbes*, published in 1576. The diagram is like all other early sketches of the Copernican universe except that the stars are no longer restricted to the surface of the celestial sphere. No stars occur within the sphere (if they did, there would be observable stellar parallax), but the infinite space beyond the sphere is studded with them. Notice, however, that the sun still retains a privileged position and that the distance between neighboring stars is far less than that between the sun and the celestial sphere. In Digges's universe the sun is not just another star.

cause no part of space could be distinguished from any other, the occupants of space — the earth, planets, and stars — must all move and must all be of the same nature.

Cusa's vision provides a second example of a cosmology that could be transformed by the existence of Copernicanism. As Cusa developed it, a hundred years before the publication of the *De Revolutionibus*, the cosmology made no scientific sense at all. In the role of cosmologist, Cusa was a mystic who joyfully rejected the appearances for the sake of a transcendent apprehension of the infinite Deity in whom all paradoxes were reconciled. Yet the Neoplatonic insistence upon the infinite and its paradoxes was not intrinsically incompatible with the appearances or with science. After Copernicus' death the same insistence provided a motive and recurrent theme for the cosmological writings of the Italian mystic, Giordano Bruno, and in Bruno's vision of the universe the infinite and the appearances were reconciled by Copernicanism. Bruno's approach to cosmology was scarcely more concerned with science or with the appearances than was the approach of Cusa, by whom he was greatly influenced. But whatever his motives Bruno was right. The sun need not be at the center; in fact, no center is needed. A Copernican solar system may be set down anywhere in an infinite universe. Providing only that the sun is far enough from the neighboring stars to account for the absence of parallax, the appearances will be preserved.

Bruno's reconciliation of an infinite and centerless universe with the appearances was only one part of his cosmological construction. Beginning around 1584 he also made explicit the physical relation of the Copernican solar system to the other celestial residents of his infinite space. The sun was, he thought, merely one of an infinite number of stars scattered through the infinite expanse of space; some of the other bodies in the infinite heavens must be populated planets like the earth. Not only the earth but the sun and the entire solar system were transformed to insignificant specks lost in the infinitude of God's creation; the compact and ordered cosmos of the scholastics had become a vast chaos; the Copernican departure from tradition had reached its maximum.

But though radical, this last extension of Copernicanism was achieved almost without novelty. Two millenniums before Bruno's birth the ancient atomists, Leucippus and Democritus, had envisaged an infinite universe containing many moving earths and many suns. In antiquity their doctrines had never rivaled Aristotle's as a basis for scientific thought, and their writings had disappeared almost entirely during the Middle Ages. But the works of their successors, Epicurus and Lucretius, were among the principal literary recoveries made by the Renaissance humanists. From these works, particularly from Lucretius' *De Rerum Natura*, Bruno derived many of his most fruitful conceptions. In Bruno's cosmology a third ancient speculative conception of the universe was revitalized and given new verisimilitude by its affinity for Copernicanism.

That affinity is somewhat surprising, for both historically and logically atomism and Copernicanism seem totally different doctrines. The ancient atomists had developed the main tenets of their cosmology not principally from observation but in an attempt to resolve apparent logical paradoxes. The existence and motion of finite bodies could, they felt, be explained only if the real world consisted of tiny indivisible corpuscles, or atoms, swimming free in a vast empty space, or void. The void was required to account for motion; if there were no empty spaces, there would be no place for matter to move into. Similarly, the indivisibility of the ultimate particles seemed to them essential for the existence of finite bodies; if matter were infinitely divisible, then its ultimate parts would be mere geometric points, occupying no space at all. From parts that, taken individually, occupy no volume it seemed impossible to construct a finite body that would. Zero plus zero equals zero no matter how often the addition is repeated. Reality must therefore, said the atomists, consist of indivisible atoms and the void, and this premise, quite foreign to Copernicanism, was the foundation of their world view.

The premise had, however, some striking consequences which were not so foreign. For example, the atomists' void must be infinite in extent. It could only be bounded by matter, and the matter must, in turn, be bounded by more void. When matter and space ceased to go together, as they had in Aristotle's physics, there was no end to the process of bounding the universe. Again, there were no special positions or unique bodies in the atomists' universe. The void itself was neutral; each position was like every other. The earth or sun existed in one region rather than another simply because the fortuitous motions and collisions of the atoms happened to produce an aggregate at that position, and because once the atoms had met by chance they became entangled and stuck. The process could equally well have occurred somewhere else. In fact, since the universe was infinite and contained infinitely many atoms, the process almost certainly had occurred elsewhere at some time. Many earths and suns as well as many atoms populated the infinite void of atomistic cosmology. There was no possible terrestrial-celestial dichotomy. According to the atomists, the same sort of matter obeyed the same set of laws everywhere in the infinite neutral void.

Since Copernicanism also destroyed the earth's uniqueness, abolished the terrestrial-celestial distinction, and suggested the infinity of the universe, the atomists' infinite void provided a natural home for Copernicus' solar system, or rather for many solar systems. Bruno's principal contribution was recognizing and elaborating this obscure affinity between the ancient and modern doctrines. Once the affinity was recognized, atomism proved the most effective and far-reaching of the several intellectual currents which, during the seventeenth century, transformed the finite Copernican cosmos into an infinite and multipopulated universe. That extension of cosmological dimensions was, however, only the first of atomism's significant roles in the construction of the new universe.

The Corpuscular Universe

Early in the seventeenth century atomism experienced an immense revival. Partly because of its significant congruence with Copernicanism and partly because it was the only developed cosmology available to replace the increasingly discredited scholastic world view, atomism was firmly merged with Copernicanism as a fundamental tenet of the "new philosophy" which directed the scientific imagination. Donne's lament that because of the "new Philosophy" the universe was "crumbled out again to his Atomies" is an early symptom of the confluence of these previously distinct intellectual currents. By 1630 most of the men who dominated research in the physical sciences showed the merger's effects. They believed that the earth was a moving planet, and they attacked the problems presented by this Copernican conception with a set of "corpuscular" premises derived from ancient atomism.

The "corpuscularism" that transformed seventeenth-century science often violated the premises of ancient atomism, but it was atomistic nonetheless. Some of the "new philosophers" believed that the ultimate particles were divisible in principle, but all agreed that they were seldom or never divided in fact. Some doubted the void, but the aethereal fluid with which they filled all space was for most purposes as neutral and inactive as the void. And, most important, all agreed that the motions, interactions, and combinations of the various particles were governed by laws imposed by God upon the corpuscles at the Creation. The discovery of these laws was, for the corpuscularian, the first problem in the program of the new science. The second was to apply these laws in explaining the rich flux of sensory experience.

The French philosopher, René Descartes (1596–1650), was the first man systematically to apply this program to the problems of a Copernican universe. He began by asking how a single corpuscle would move in the void. Then he asked how this free motion would be altered by collision with a second corpuscle. Since he believed that all change in the corpuscular universe resulted from a succession of free corpuscular motions punctuated by intercorpuscular collisions, Descartes expected to deduce the entire structure of the Copernican universe from the answers to a few questions like these. Though all of his deductions were intuitive and though most of them were mistaken, the cosmology that his imagination dictated to his reason proved immensely plausible. Descartes's vision dominated much of science for almost a century after its details were first published in his *Principles of Philosophy* in 1644.

Descartes's answer to his first question was extremely successful. By applying contemporary. versions of the medieval impetus theory to a corpuscle in the infinite neutral space of atomistic cosmology, he arrived at the first clear statement of the law of inertial motion. A corpuscle at rest in the void will, he said, remain at rest, and a corpuscle in motion will continue to move at the same speed in a straight line unless it is deflected by another corpuscle. The constancy of particle speed was a straightforward consequence of the impetus theory (discussed in the penultimate section of Chapter 4), particularly as that theory had been developed by Galileo. But the linearity of motion was a novelty and an immensely consequential one, typifying the fruitful suggestions which atomism offered to seventeenth-century science. The atomists' infinite void was a space with no center and (except in a few debased versions which were discarded early in the century) without any intrinsic directions, like "up" and "down." In such a space a body free from external influence could only stop or move straight ahead. The self-sustaining circular motions borrowed from the scholastic impetus theory by Copernicus, Galileo, and other early Copernicans were impossible. After Descartes's work they played no significant role in the construction of the Copernican universe.

In nature, Descartes recognized, all particles or aggregates of particles continually change their speed and direction. These alterations must, said Descartes, be caused by external pushes and pulls derived from other bodies (Figure 46). Corpuscular collisions were therefore

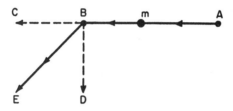

Figure 46. The effect of an impulse upon an inertial motion. At the point *A* the body *m* is given a sharp push toward *B*. If no additional pushes intervene, the body will move straight from *A* toward *B* at a constant speed. If the body is given a second push in the same direction when it reaches *B*, it will continue straight on toward *C* but at an increased speed. If pushed in the opposite direction, it may continue toward *C* but at reduced speed, or, if the second push is strong enough, the body may recoil from *B* and move straight back toward *A*. Finally, if the body when it reaches *B* is given a lateral push toward *D*, it will begin a new inertial motion along the slant line *BE*. The inertial motion along *BE* may be thought of as the result of two simultaneous inertial motions, one along *BC*, generated by the first push at *A*, and the other along *BD*, generated by the second lateral push at *B*.

the second subject for investigation, and with it Descartes was less successful. Only one of his seven laws of collision was retained by his successors. But though Descartes's laws of collision were discarded, his conception of the collision process was not. Corpuscularism had again introduced a new problem, and that problem was solved within thirty years of Descartes's death. From the solution came both the law of conservation of momentum and, though less directly, the conceptual relation between a force and the change of momentum that it produces. Both were essential steps toward Newton's universe.

In proceeding from his laws of motion and collision to the structure of a Copernican universe, Descartes introduced a concept which since the seventeenth century has greatly obscured the corpuscular basis of his science and cosmology. He made the universe full. But the matter that filled Cartesian space was everywhere particulate in structure, and in determining the behavior of this particulate plenum Descartes made constant imaginative use of the void. He used it first to determine the laws of motion and collision for individual particles. Then, to discover how these laws operated in a plenum, he seems first to have imagined the particles swimming in a void where their inertial motions were punctuated by collisions; after which he gradually squeezed the void out of the system, bringing the particles closer and closer together, until finally their collisions and inertial motions merged into a single process in the plenum. Unfortunately, in the plenum the motions of all particles must be considered simultaneously, an incredibly complex problem which Descartes scarcely tried to solve. Instead he indulged in an imaginative leap from his corpuscular laws to his final solution, pausing for none of the absolutely essential intermediate steps.

To Descartes it seemed self-evident that the only enduring motions in a plenum must occur in circulatory streams. Each particle in such a stream pushes forward its nearest neighbor until, finally, to avoid a vacuum, the push is returned to the first particle over an approximately circular path. Then, having filled the potential void, the process starts again. Since, for Descartes, these circulatory streams were the only possible enduring motions, he believed that, whatever push God gave to the corpuscles at the creation, they would ultimately circulate in a set of vortices scattered through all of space. A small set of these vortices is reproduced in Figure 47 from an illustration for one of Descartes's early works.

Each of Descartes's vortices was, at least potentially, a solar system, generated and governed by the corpuscular laws of inertia and collision. For example, corpuscular impacts just balance the centrifugal tendency that inertia gives to each corpuscle in the vortex. If all others were removed, each single particle would travel straight ahead along a tangent to its normally circular path and thus leave the

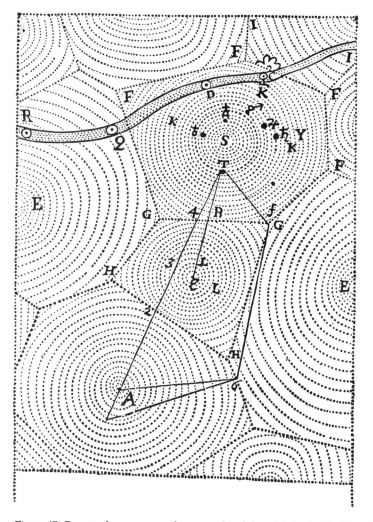

Figure 47. Descartes's vortex-cosmology reproduced from his book, *The World or a Treatise on Light*. The points S, E, A, and ε are the centers of vortices. The rapid churning motion of the restricted central corpuscles makes these centers self-luminous, so that they act like stars. The various dotted circles, which need not be precisely circular, represent the paths of the eternally rotating corpuscular streams that compose the vortex. The dots around the vortex center S are planets, swept through their orbits by the vortical flow. The body C, which crosses the top of the diagram, is a comet, passing from vortex to vortex in a region where the flow is too slow to trap it in a continuing circular orbit. Other vortices fill the space outside the diagram, and each one represents, at least potentially, the location of a solar system in Descartes's multipopulated Copernican universe.

vortex. It does not do so only because constant collisions with particles outside of it in the vortex continually drive it back toward the center. Similar impacts keep the stable corpuscular aggregates that form the planets circulating in approximately circular paths about the vortex center.

The constant rapid churning motion of each vortex center sets up a continual vibration which is transmitted in waves from the center through all of space. According to Descartes that vibration is light, continually emerging from the suns or stars which are the vortex centers. Apparently an infinite multiplicity of star-centered planetary systems have been derived from corpuscular premises. And Descartes's derivations do not stop with these celestial phenomena. For example, he explains the motion of the moon, the tides, and projectiles by positing a set of small subsidiary vortices, one around each planet. Corpuscular impacts within these smaller vortices keep the moon in circulation and drive falling stones to earth. In Descartes's universe weight itself, like motion, light, and all other sensory appearances, is ultimately traceable to corpuscular collisions governed by the laws of atomic motion and interaction.

It is, today, childishly easy to discover errors and inadequacies in Descartes's discussion of vortex cosmology and in the astronomy, optics, chemistry, physiology, geology, and dynamics that he derived from it. His vision was inspired, and its scope was tremendous, but the amount of critical thinking devoted to any one of its parts was negligibly small. His laws of corpuscular collision provide but one of countless examples. But in the development of seventeenth-century science the parts of Descartes's system were far less important than the whole. Descartes's brilliant successors, led by Christian Huyghens, found their inspiration in his underlying conception, rather than in its detailed development. They could and did change his laws of collision, his description of vortices, and his laws for the propagation of light. But they did not compromise his conception of the universe as a corpuscular machine governed by a few specified corpuscular laws. For half a century that conception guided the search for a self-consistent Copernican universe. It seems no coincidence that this basic conception of the structure appropriate to a Copernican universe was so largely inspired by an ancient world view which Copernicanism itself had helped make popular.

The Mechanical Solar System

Two quite separate historical paths lead from Copernicus' finite sun-centered cosmos to the Newtonian universe that gave the Copernican Revolution its final form. One of these is the liaison, illustrated above, between Copernicanism and the corpuscular philosophy. The other is a narrowly focused series of attacks upon the single most pressing physical problem raised by Copernicanism: What moves the planets? Both paths began half a century after Copernicus' death. Their common source is the new scientific perspective created when Kepler, Bruno, and others separated the truly novel from the pseudo Aristotelian elements in Copernicus' work. The two paths merge again in Newton's terminal formulation of the structure of the Copernican universe, and both provided essential elements to that formulation. But except at their initial and final termini the two paths were most often separate, though striking parallelisms gave occasional evidence that they were proceeding in the same direction.

The physical explanation of planetary motion was not, in the sixteenth and seventeenth centuries, quite a new problem. Neither Aristotle, Ptolemy, nor the medieval astronomers had been able fully to specify the physical cause of each minor irregularity in a planet's motion. But traditional science had at least explained the average eastward drift of all planets around the ecliptic. Planets and the spheres in which they were embedded were made from a perfect celestial element whose nature expressed itself in eternal rotations about the center of the universe.

Copernicus had tried to preserve this traditional explanation of planetary motion. But the conception of natural celestial motions was less suitable to a sun-centered than to an earth-centered universe, and the incongruities of Copernicus' initial proposal did not remain hidden for long. To explain even the planets' eastward drift, the Copernican system demanded that each particle of earth rotate naturally about two different centers — the fixed center of the universe and the moving center of the earth. Each particle of a satellite, like the moon, was simultaneously governed by at least three centers — the center of the universe, the center of the governing planet, and the center of the satellite itself. Copernicanism therefore jeopardized the plausibility of self-sustaining circular motions by compounding them and by re-

lating them to many simultaneous fixed and moving centers. Furthermore, the multiplicity and motion of the various centers deprived Copernican motions of any fixed relation to the intrinsic geometry of
space. In Aristotelian physics all natural motions had been either
toward, or away from, or about the center of the universe. That center,
though only a geometric point, could plausibly possess a special causal
role because it was unique, determined once and for all by its relation
to the boundary of space. Copernicus' proposal, on the other hand,
demanded that some natural motions be governed by moving centers,
and moving centers could not act causally by virtue of geometric position alone.

Late in the sixteenth and early in the seventeenth century other
new astronomical doctrines converged to make the physical problem
of the planets still more acute. All the celestial spheres except the
sphere of the stars were made obsolete by the new observations of
comets and by the increasing popularity of the Tychonic system. With
the spheres went the entire physical mechanism that had previously
accounted for the planets' average circular motions. Even this dissolution of spheres did not end the influence of the classical approach.
As late as 1632 Galileo could still attempt to elaborate Copernicus'
physical doctrine, arguing in his *Dialogue on the Two Principal Systems of the World* that even without celestial spheres all matter would
rotate naturally, regularly, and eternally in a set of compounded circles. But the brilliance and subtlety of Galileo's dialectic — scarcely
rivaled since in a major scientific work — could not for long disguise
the fundamental implausibility of the approach. His *Dialogue* was
important as a great popularization of Copernicanism, but his monu-
mental contributions to physical science are in other works. After his
death progress on the physical problem of the planets took a quite
different direction, because even before Galileo's *Dialogue* appeared
Kepler's researches had given the physical problems of Copernicanism
a new dimension and suggested a new set of techniques for their
solution.

By doing away with the profusion of epicycles and eccentrics, Kepler made it possible, for the first time, to subject the full complexity
of the celestial appearances to physical analysis. An explanation which,
like Copernicus' or Galileo's, dealt only with the average eastward
drift of planets ceased to seem adequate even when plausible. The

geometrically simple and precise elliptical motions rather than the average drifts now demanded explanation. But the new precision and simplicity were achieved only at a price. In contrast to the average circular motions of classical astronomy, elliptical motions governed by the Second Law could not be natural motions, for they were not symmetric with respect to any center. A planet moving uniformly on a deferent, or even on a simple epicycle-deferent system, is in some sense "doing the same thing" or "moving in the same way" at each and every point in its orbit; this motion might conceivably be "natural" to it. On the other hand, the motion of a planet obeying Kepler's Laws changes speed, direction, and curvature at each point in the orbit. These variations seemed to demand the introduction of a force in the heavens, a force acting continually to alter the planet's motion at each point in its orbit. In the heavens as on the earth an asymmetric motion was most naturally explained as the result of a continuing push or pull.

In other words, Copernicus' innovation first destroyed the traditional explanation of planetary motion and then, as modified by Kepler, suggested a radically new approach to celestial physics. That new approach first appeared in the writings of Kepler himself during the last decades of the sixteenth century and the first decades of the seventeenth. In essence it was an inversion of the technique which Copernicus had already used, and which Galileo was to resurrect, in unifying terrestrial and celestial laws. Copernicus and Galileo achieved uniformity by applying the traditional conception of natural circular celestial motions to the earth. Kepler achieved the same effect more fruitfully by applying the ancient conception of violent force-governed terrestrial motions to the heavens. Guided by his ever-present Neoplatonic perception of the sun, Kepler introduced forces emanating from the sun and planets to provide a causal foundation for planetary motion. In his writings the solar system was for the first time modeled on a terrestrial machine. For all the crudities of its initial conception, the future lay with Kepler's approach.

The first of Kepler's solar forces was the *anima motrix*, which we examined briefly in Chapter 6. Kepler visualized it as a system of rays projecting from the sun in the plane of the ecliptic and carried about by the sun's continuing rotation. As the moving arms passed a planet, they pushed against it, impelling it in a continuing circle about the sun. To change the basic circular orbit to an ellipse, a second force was required, one that would change the distance between sun and planet in different portions of the orbit. That second force Kepler identified as magnetism, whose properties had recently been thoroughly investigated and recorded in a very influential book, *On the Magnet*, published in 1600 by the English physician William Gilbert. Gilbert had recognized that the earth itself was a huge magnet, and Kepler extended the generalization to the other bodies of the solar system. Not only the earth, said Kepler, but also the planets and the sun are magnets, and the attractions and repulsions of their various poles determine the paths in which the planets move.

Few of Kepler's successors took his physical theory, whose details are indicated in Figure 48, as seriously as they took his mathematical description of the planetary orbits. Some of his dynamical concepts

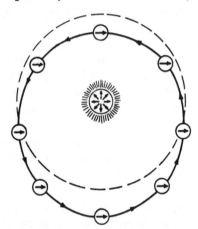

Figure 48. Kepler's mechanical solar system. The sun is shown at the center of the diagram, and the lines radiating from it represent the *anima motrix* which, in the absence of additional magnetic forces, would push the planet, *P*, around the broken circle centered on the sun. The magnets which, according to Kepler, transform this circular motion into an ellipse are represented by arrows. The sun's south pole is buried at its center, where the pole exerts no influence. Its north pole is distributed uniformly over the sun's surface. The earth's magnetic axis always remains very nearly parallel to itself as the earth moves. Therefore, whenever the planet is to the right of a central vertical line through the diagram, its south pole is nearer the sun than its north, and it is attracted gradually toward the sun. During the other half of its motion, the planet is gradually repelled. At all positions the planet's orbital speed is inversely proportional to its distance from the sun, because the *anima motrix* is stronger near the sun.

were out of date when he wrote; the sun rotates too slowly to account for the observed periods of the planets; when observed with magnetic needles the direction of the earth's magnetic axis is different from that required to account for astronomical observations. Therefore, after Kepler's death neither the *anima motrix* nor the magnetic sun often appeared in seventeenth-century scientific writings. But Kepler's conception of the solar system as a self-contained and self-governing machine recurred again and again. As Copernicanism developed during the seventeenth century, that conception proved to have a double importance.

In the first place, Kepler's physical system, though entirely independent of the corpuscular philosophy, reinforced some of corpuscularism's most significant conclusions. In particular it provided a second natural route to the conception of an infinite neutral space. In Kepler's planetary mechanism a planet's motion depended only on its relation to another physical body, the sun. Magnetism and the *anima motrix* functioned equally well whatever the sun's position. Though Kepler himself kept the sun at the center of a finite stellar sphere, no such center was necessary. Corpuscularism had reached similar conclusions, but from quite different motives and by a totally different route. Apparently some of Copernicanism's most striking consequences could not be suppressed by any of the approaches that led to a coherent Copernican universe.

Kepler's substitution of violent, force-determined planetary motions for the natural, force-free, and space-determined motions of traditional celestial physics has a second major role in the development of seventeenth-century science. Kepler's mechanical solar system is the first in a series that culminates in the system of Newton's *Principia*. Historically, the intervening developments are extremely complex. They depend upon the tortuous evolution and laborious assimilation of a new set of dynamical concepts and mathematical techniques, developments that by themselves could be the subject of another book. But conceptually the route from Kepler to Newton is relatively simple. A few significant emendations will convert Kepler's system to one that is qualitatively very like Newton's, and these emendations are direct consequences of recognizing the role in celestial physics of Descartes's conception of inertial motion. The absence of that conception is the principal feature which distinguishes Kepler's mechanical solar system from the similar systems designed by Newton's immediate predecessors. Two of these later systems, designed by the Italian G. A. Borelli (1608–1679) and by the Englishman Robert Hooke (1635–1703), will bring us very close to the qualitative features of Newton's system.

Borelli's conception of inertial motion was far less complete than Hooke's, and his planetary theory was therefore far closer to Kepler's. Unlike Kepler, Borelli realized that no push was necessary simply to keep the planets from coming to rest. But he retained a sort of *anima motrix* to account for the variation of a planet's speed with its distance from the sun, and on occasions he too seems to have thought of the *anima motrix* as a continuing pusher. Elsewhere his break with Kep-

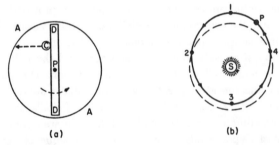

(a)   (b)

Figure 49. Borelli's planetary theory. Diagram (*a*) shows Borelli's model of the planetary mechanism in which a cork *C* is driven counterclockwise around the bowl *AA* by the rotating arm *DD*. When the arm revolves rapidly, the cork spirals quickly out to the rim of the bowl under the dominant influence of its inertial tendency to move in a straight line. If the arm moves slowly, the cork spirals toward the center of the bowl because the slight centrifugal tendency generated by the arm's slow revolution is more than overcome by the attraction between the magnets mounted at *C* and *P*. At an appropriate intermediate speed the two opposing tendencies, centripetal and centrifugal, will just balance, and the cork will move in a continuous circle, the basic Copernican orbit.

Diagram (*b*) illustrates Borelli's derivation of an elliptical orbit. When the planet moves around the broken circle, the centrifugal tendency generated by the *anima motrix* just balances the planet's tendency to fall into the sun, and the orbit is therefore a circle. If the planet is now removed from the broken circle and placed on the solid curve at position 1 outside the circle, its motion will be slower, corresponding in diagram (*a*) to a slower rotation of the arm, because at a greater distance from the sun the *anima motrix* is weaker. As a result the planet will begin to spiral inward toward the sun along the solid line, crossing the broken circle at 2, and continuing to 3 where the increased strength of the *anima motrix* has increased its speed sufficiently to overcome the inward drift. Then the planet begins an outward spiral which carries it back to 1. Borelli claimed that the resulting orbit would be an ellipse.

ler (and Aristotle) was more complete. In particular, Borelli recognized (and demonstrated with an imagined model described in Figure 49) that no push like that from the *anima motrix* could keep a planet moving in a closed orbit. Unless some other force were available to pull the planets straight toward the sun, each of them would, Borelli thought, move off along a straight line tangent to its orbit and thus leave the solar system entirely. To maintain stable orbits Borelli therefore introduced a second force which constantly deflected the departing planet back toward the sun. In his model Borelli used magnets to simulate this force; in the heavens he displayed the residual strength of Aristotelian concepts by replacing the force with a natural tendency of all planets to fall toward the central sun.

Borelli's conception of the solar system was elaborated in a book published in 1666, the same year in which Robert Hooke at last demonstrated the full parallelism between the celestial motions and those of a terrestrial machine. Much influenced by Descartes, Hooke began with a complete conception of inertial motion and of the identity of terrestrial and celestial laws. As a result he was able to discard both the *anima motrix* and the vestiges of natural tendencies to motion. A moving planet ought, he said, to continue its motion uniformly in a straight line through space, because the senses reveal nothing to push or pull it. Since its motion is not straight, but rather in a continuous closed curve encircling the sun, the immediate evidence of the senses must be misleading. There must be an additional attractive principle or force operating between the sun and each planet. Such a force would, said Hooke, continually deflect the planets' straight inertial motions toward the sun, and that is all that their Copernican orbits require.

Hooke's intuitive perception of a Copernican planetary motion is indicated by Figure 50a, though in a form more explicit than any provided by Hooke himself. The solid circle (or it might be an ellipse) is the planet's Copernican orbit, and the planet is shown in this orbit at *P*, where it is moving with a constant speed. If there were no force between the sun, *S*, and the planet, then the planet should move straight out along the broken line tangent to the orbit, always at the same speed. But if, when the planet is at *P*, it is suddenly pushed sharply toward the sun, then (remember Figure 46) it will acquire a simultaneous motion toward the sun, indicated by the short broken radial line in the diagram. The result of these two motions will be a new inertial motion along the solid arrow in the diagram, rejoining the actual orbit at P'. If the planet at P' were again pushed sharply toward the sun, it would begin to move along the second solid arrow, toward P'', and the process could be continued until the planet finally returned to P.

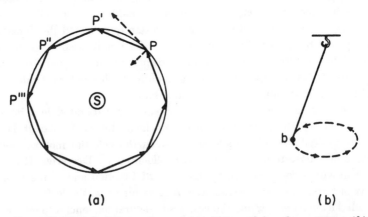

(a)   (b)

Figure 50. Hooke's planetary theory (a) and his pendulum demonstration (b). In diagram (a) the planet is given a sudden push toward the sun, S, at each of the points P, P', P'', ... Each push changes the direction of the planet's inertial motion, and the result of all the pushes is motion along the perimeter of a polygon. Increasing the number of points at which the planet is pushed toward the center increases the number of sides of the polygon. In the limit a continuous push toward the center produces motion around a circle. This circular motion is demonstrated physically with the apparatus sketched in diagram (b). Unless given a lateral push the pendulum bob, b, will simply be drawn by its weight toward a point close to the center of the dotted circle. But when the bob is given an initial horizontal push in a direction perpendicular to the string, the weight of the bob can only deflect the motion into a curve. If the push is of the right strength, the bob will revolve in a horizontal circle or in an elongated orbit that closely approximates an ellipse.

The succession of pushes described above does not move the planet around the smooth curve representing its orbit. Instead the planet moves around a polygon. But the polygonal series of solid arrows approximates the planet's orbit, and the approximation can be improved indefinitely. Suppose, for example, that the magnitudes of the impulses delivered at P, P', P'', . . . are reduced so that the planet is deflected less at each of these points and therefore rejoins its curved orbit sooner; and suppose that the original series of impulses (now reduced in strength) are supplemented by a new series delivered at the points between P and P', P' and P'', . . . where the planet now rejoins the curve. The resulting motion will still describe a polygon rather than an ellipse or circle, but the polygon now approximates the circle more closely. As the individual impulses are further decreased in strength and increased in number, the approximation becomes still closer. Finally, when the individual impulses become infinitely small and infinitely numerous, the planet is deflected toward the sun at each point in its trajectory, and if the deflecting force is always of the proper strength, the resulting curve will be just the desired ellipse or circle.

This was Hooke's hypothesis, and it remained quite vague. Hooke did not know how to relate the magnitude of a force to the size of the deflection it would produce, and he did not know how to generate an ellipse from a continuous series of deflections. He did not and could not show that his hypothesis would work. That job was left for Newton. But Hooke was able to give his idea a concrete and plausible form by setting up a model that produced motions like the planetary motions under the influence of a single centrally directed force. In 1666 he concluded the lecture whose content we have just sketched by showing his colleagues at the Royal Society a so-called conical pendulum (Figure 50*b*), constructed by suspending a heavy bob from a wire free to move in any direction. When the bob was pulled slightly aside from its lowest position, the only effective force exerted on it was a pull directed approximately toward the low point of the pendulum, one wire length under the point of suspension. Released from a position of rest away from this low point, the bob was simply pulled back, and it oscillated steadily to and fro in a plane like an ordinary pendulum. But when the bob, after being displaced, was given a sharp horizontal push in a direction perpendicular to the line connecting it with the low point, then it did not necessarily return to its lowest point at all. Instead it could swing steadily in a horizontal plane about the low point, executing a continuous orbit like that of a planet. When started in the appropriate direction with the appropriate speed, the bob revolved in a horizontal circle. With a slightly different initial speed it moved in an elongated orbit, quite like an ellipse. The centrally directed force was unable to pull the moving bob to the center.

It only deflected the motion toward the center, thus producing a continuous curve. A single centrally directed force had produced in the laboratory a closed orbit of the proper shape. A similar force in the heavens should, said Hooke, have the same effect.

Hooke's model made his vaguely enunciated theory both clear and plausible, but it has a larger importance as well. To us the model provides a forceful illustration of the immense and fruitful change that the physical problem of the planets had undergone through the influence first of Copernicanism and then of corpuscularism and Copernicanism combined. In Hooke's work, even more than in Kepler's and Borelli's, the explanation of planetary motion has become a problem in applied mechanics, identical in principle with the terrestrial problems of the pendulum and the projectile. Terrestrial experiments yield direct knowledge of the heavens, and celestial observations give information immediately applicable on the earth. The breakdown of the terrestrial-celestial dichotomy, demanded by the *De Revolutionibus* and facilitated by the corpuscular philosophy, is at last complete. Crystal spheres and all other special celestial devices have been banished and replaced by a mechanism of terrestrial type, and that mechanism has been shown to function as adequately as Aristotle's spheres.

Gravity and the Corpuscular Universe

One other pressing problem raised by Copernicus' innovation played an essential role in the evolution of the new universe: Why do heavy bodies fall to the surface of a moving earth whatever the earth's position in space? Though philosophers have maintained that scientists should not ask questions of this sort — questions beginning with "Why" — they were certainly asked, and asked fruitfully, during the seventeenth century. Descartes, for example, provided one answer to this particular question — loose bodies are driven to earth by the impact of aerial corpuscles in the earth-centered vortex — and that answer was widely believed until after Newton's death. A competing solution had, however, been developed by earlier Copernicans: heavy bodies are drawn to earth by an intrinsic attractive principle which acts between all pieces of matter. Once modified to fit at least a few of corpuscularism's major premises, the answer based on an intrinsic principle of attraction triumphed over the purely corpuscular explanation of Descartes and his followers. By the end of the century that attractive principle, known today as gravity, had provided the key to most motions on the earth and all those in the heavens.

Like most conceptions current in seventeenth-century science, gravity has precursors reaching back to antiquity. Some of Plato's predecessors, for example, had believed that similar substances must attract or, alternatively, repel each other. But, except in the study of magnetism and electricity, these innate attractive and repulsive principles had had few concrete applications until the concept of a planetary earth called them forth. The obscure connection between these two apparently disparate concepts, gravity and the planetary earth, appears clearly and early in a passage we have already noted from Oresme's commentary on Aristotle's *On the Heavens* (p. 114). There could, said Oresme, be many earths in space, but in that case stones must fall to earth because matter aggregates naturally to matter rather than because it aggregates to the geometric center of the universe.

In the *De Revolutionibus*, Book One, a similar need called forth a conception very like Oresme's. "Now it seems to me," said Copernicus, "gravity [which here means simply weight] is but a natural inclination, bestowed on the parts of bodies by the Creator so as to combine the parts in the form of a sphere" (see p. 153 above). Kepler, too, elaborated the idea of an attractive principle acting between the earth and its parts. He even suggested that the same principle might act reciprocally between the earth and the moon. Only in considering bodies outside of the earth-moon system did Kepler feel the need of special celestial forces like the *anima motrix*. Until Descartes's corpuscular explanation of weight was published in 1644, most Copernicans continued to explain the descent of stones by some device like Kepler's. Either there was an intrinsic attractive principle, like magnetism, by which the earth attracted stones and stones attracted the earth, or else (the equivalent only for present purposes) stones possessed an intrinsic tendency to move toward the physical center of the earth.

After the middle of the century these Copernican explanations of the fall of a stone were rapidly applied to the new Copernican problem raised by the assimilation of the concept of inertial motion. First Descartes and then Borelli, Hooke, Huyghens, and Newton, all recognized that to pursue a closed orbit about the sun a planet must continually "fall" toward the sun, thus transforming its linear inertial motion into a curve. As the need for an explanation of this "fall" was recognized, each Copernican adapted some variant of his explanation for terrestrial fall to the celestial case. Descartes's planets were pushed toward the sun by corpuscular impact; Borelli's possessed a natural tendency to move toward the sun; and Hooke's were drawn to the sun by an intrinsic mutual attraction. This much we have already seen.

Hooke, however, and Newton at about the same time, made one other immensely consequential step. Led, perhaps, by Descartes's idea that the same mechanism governed terrestrial and celestial fall, they suggested that the force which drew planets to the sun and the moon to earth was the same gravitational attraction which caused the fall of stones and apples. We shall probably never know which of the two reached the conception first. But Hooke was at least the first to announce it publicly, and his statement of 1674 is still worth reading as a clear description of the vision which, made quantitative and corpuscular by Newton, guided the scientific imagination of the eighteenth and nineteenth centuries. Hooke wrote:

[At a future date] I shall explain a System of the World differing in many particulars from any yet known, [and] answering in all things to the common rules of mechanical motions. This depends upon three suppositions: first, that all celestial bodies whatsoever have an attraction or gravitating power towards their own centers, whereby they attract not only their own parts, and keep them from flying from them, as we may observe the earth to do, but that they do also attract all the other celestial bodies that are within the sphere of their activity; and consequently that not only the sun and moon have an influence upon the body and motion of the earth, and the earth upon them, but that Mercury, also Venus, Mars, Jupiter, and Saturn, by their attractive powers, have a considerable influence upon its motion as in the same manner the corresponding attractive power of the earth hath a considerable influence upon every one of their motions also. The second supposition is this: that all bodies whatsoever that are put into a direct and simple motion, will so continue to move forward in a straight line, till they are by some other effectual powers deflected and bent into a motion, describing a circle, ellipse, or some other more compounded curve line. The third supposition is: that these attractive powers are so much the more powerful in operating, by how much the nearer the body wrought upon is to their own centers. Now what these several degrees are I have not yet experimentally verified; but it is a notion, which if fully prosecuted as it ought to be will mightily assist the astronomer to reduce all the celestial motions to a certain rule, which I doubt will never be done true without it.[1]

Hooke's first two "suppositions" are fundamental premises of the new universe. Inertia plus a single attractive force, gravity, governs both celestial motions and the motions of terrestrial projectiles. By implication, at least, planets and satellites are simply terrestrial projectiles, cannon balls fired with a muzzle velocity so great that they never quite fall to the earth's surface but instead circle it continually. Newton himself made this image explicit and familiar in his *System of the World* (Figure 51). But Hooke's remarks supply something more

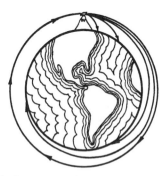

Figure 51. Newton's description of the cannon ball as a satellite. As the velocity of the ball is increased the length of its trajectory also increases, so that it travels farther around the curved surface of the earth. When the velocity is great enough, the ball does not fall to earth at all, but instead revolves continually in an approximately circular orbit.

than conceptual foundations. The passage just quoted also makes explicit two of the main problems that remained to be solved before the new universe would be complete. How does the gravitational force vary with the distance between the attracting bodies, and how can a knowledge of this law of attraction be used to predict both terrestrial and celestial motions?

With these problems Hooke himself could do nothing. He was not enough of a mathematician to deduce the law of attraction from Kepler's description of the planetary orbits; the instruments that he carried to the top of St. Paul's cathedral and to the bottoms of mines were far too insensitive to detect the small variation of gravity near the earth's surface. But Hooke was not the only scientist working in the field. Though neither he nor his other contemporaries knew it, Isaac Newton (1642–1727) had already arrived independently at an important part of Hooke's qualitative conception. Furthermore, if his own subsequent dating of the discovery is reliable, Newton had used the conceptions to determine Hooke's "several degrees" of gravitational attraction eight years before the passage above was written.

When Newton turned to the problem around 1666, he succeeded in working out mathematically the rate at which a planet must "fall" toward the sun, or the moon toward the earth, in order to remain in a stable circular orbit. Then, having discovered how this mathematical rate of fall varied with the planet's speed and with the radius of its circular orbit, Newton was able to deduce two immensely important physical consequences. If the speeds of the planets and their orbital radii were related to each other by Kepler's Third Law, then the attraction that drew planets to the sun must, Newton found, decrease inversely as the square of the distance separating them from the sun. A planet twice as far from the sun would require only one-fourth the attractive force to remain in its circular orbit at its observed speed. Newton's second discovery was equally far-reaching. The same inverse square law that governed the attraction between sun and planets would, he found, account quite well for the difference in the rate at which the distant moon and a nearby stone fell to the earth. Thirteen years later, recalled to the problem by a controversy with Hooke, he generalized his results still further and showed that an inverse-square law would account precisely for both the elliptical orbits specified by Kepler's First Law and the speed variation described in the Second.

These mathematical derivations were without precedent in the history of science. They transcend all the other achievements that stem from the new perspective introduced by Copernicanism. Our inability to pursue them in an elementary treatise produces the single worst distortion in this foreshortened epilogue to the Copernican Revolution. From Newton's inverse-square law and the mathematical techniques that related it to motion, both the shape and the speed of celestial and terrestrial trajectories could be computed for the first time with immense precision. The resemblance of cannon ball, earth, moon, and planet was now seen, not in a vision, but in numbers and measurement. With this achievement seventeenth-century science reached its climax. Yet strangely enough that climax was not quite the end of the Copernican Revolution. Despite its scope and power, neither Newton nor many of his contemporaries were satisfied with the concept of gravity and its operation. By 1670 the corpuscular philosophy provided the metaphysical background for almost all progressive research, and the concept of gravity violated corpuscular premises in two significant respects. Another half-century of research and argumentation were required for the reconciliation. In the new universe that finally emerged both corpuscularism and Newton's concept of gravity had been altered once more.

Newton, whose continuing allegiance to corpuscularism is repeatedly attested by his letters and college notebooks, was himself intensely aware of the metaphysical inadequacy of his working concept of gravity. That awareness probably accounts for at least part of his delay in announcing the results of his early work in celestial physics. In fact, the *Principia* did not appear until Newton, in 1685, succeeded in resolving one of the apparent conflicts between gravity and the corpuscular philosophy and until he had expended much fruitless effort in attempting to resolve the other.

The first conflict between corpuscular premises and Newton's early theory of gravity appears in the calculation of 1666, which compared the earth's attraction for the distant moon and a nearby stone. By comparing the stone's rate of fall with the moon's, Newton concluded that the earth's attraction for a unit mass outside its surface varied inversely as the square of the distance between the exterior mass and *the earth's center*. The conception was simple and in adequate agreement with experiment. Furthermore it applied brilliantly to the entire solar system. But it was not corpuscular. According to a corpuscularian the earth's attraction for an external corpuscle can be discovered only by adding together the attractions that each individual corpuscle within the earth exerts upon the single exterior corpuscle (Figure 52). If the exterior corpuscle is very distant from the earth, the addition is easy. In that case the exterior corpuscle is approximately equidistant from each corpuscle in the earth; each earth corpuscle, wherever located, exerts approximately the same force on the exterior corpuscle; and the total force must be very nearly the same as it would be if all the earth corpuscles were moved slightly to their mean position and concentrated together at the earth's center. Therefore, if the attraction of individual particles is governed by the inverse-square law, the attraction of gross bodies at large distances must be governed by the same law.

The addition of microscopic forces is not, however, nearly so easy when the exterior corpuscle is close to the earth's crust. In this case the gross inverse-square law seems most unlikely to apply. When close to the earth (Figure 52), the external corpuscle is millions of times nearer to the nearby earth corpuscles than to those on the opposite side of the earth. The nearby corpuscles therefore exert an immensely stronger force than the more distant ones. Apparently they will exert almost all the force, and the total attraction will increase extremely rapidly as the external corpuscle draws close to the earth's surface. The distance

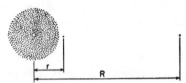

Figure 52. Making the inverse-square law corpuscular. If gravity is a corpuscular attraction, then the total attraction of the earth for an exterior corpuscle must be just the sum of the attractions between each earth corpuscle and the external corpuscle. It is by no means clear that this total attraction varies in any simple way with distance. Newton, however, succeeded in showing that if the attraction between individual corpuscles varied inversely as the square of the distance between them, then the total attraction between the earth and an exterior corpuscle would vary inversely as the square of the distance from the earth's center to the exterior corpuscle.

to the earth's *center* seems almost irrelevant in calculating the total force exerted on, say, an apple. Newton was able to show that it is not irrelevant. In 1685 he proved that, whatever the distance to the external corpuscle, all the earth corpuscles could be treated as though they were located at the earth's center. That surprising discovery, which at last rooted gravity in the individual corpuscles, was the prelude and perhaps the prerequisite to the publication of the *Principia*. At last it could be shown that both Kepler's Law and the motion of a projectile could be explained as the result of an innate attraction between the fundamental corpuscles of which the world machine was constructed.

Yet even this corpuscular conception of gravity did not satisfy Newton. Indeed, until the eighteenth century, it satisfied very few scientists. To most seventeenth-century corpuscularians gravity, as an innate attractive principle, seemed far too much like the Aristotelian "tendencies to motion" which they were unanimous in rejecting. The great virtue of Descartes's system had been its complete elimination of all such "occult qualities." Descartes's corpuscles had been totally neutral; weight itself had been explained as the result of impact; the conception of a built-in attractive principle operating at a distance therefore seemed a regression to the mystic "sympathies" and "potencies" for which medieval science had been so ridiculed. Newton himself entirely agreed. He repeatedly attempted to discover a mechanical explanation of the attraction, and though forced at last to admit his failure, he continued to maintain that someone else would succeed, that the cause of gravity was not "uncapable of being discovered and made manifest." [2] Again and again he insisted that gravity was not innate in matter. "To tell us," he proclaimed in his scientific testament at the end of his *Opticks*, "that every Species of Things is endow'd with an occult specifick Quality [like Gravity] by which it acts and produces manifest Effects, is to tell us nothing." [3]

It does not, I think, misrepresent Newton's intentions as a scientist to maintain that he wished to write a *Principles of Philosophy*, like Descartes, but that his inability to explain gravity forced him to restrict his subject to the *Mathematical Principles of Natural Philosophy*. Both the similarity and the difference of titles are significant. Newton seems to have considered his magnum opus, the *Principia*, incomplete. It contained only a mathematical description of gravity. Unlike Descartes's *Principles* it did not even pretend to explain why the universe runs as it does. It did not, that is, explain gravity, or so Newton thought. But though twentieth-century science has justified Newton's misgivings — gravity can be explained without recourse to an innate attractive principle that acts at a distance — few of Newton's contemporaries and successors were willing to preserve his subtle distinctions. Either they rejected the whole idea of gravity as reversion to Aristotelianism, or they accepted the concept and insisted that Newton had shown gravity to be an intrinsic property of matter.

The resulting battle was anything but trivial. It was forty years before Newtonian physics firmly supplanted Cartesian physics, even in British universities. A few of the ablest physicists of the eighteenth century continued to seek a mechanical-corpuscular explanation of gravity. But no such explanation was found; meanwhile the power of the *Principia* made it indispensable to scientists. Gravity was, therefore, gradually accepted, and despite Newton's disclaimers it became an intrinsic property of the ultimate corpuscles of matter.

As a result the corpuscular philosophy was remade and the search for forces began. Near the beginning of the *Principia* Newton had said,

I am induced by many reasons to suspect that . . . [the phenomena of Nature] may all depend upon certain forces by which the particles of bodies, by some causes hitherto unknown, are either mutually impelled towards one another, and cohere in regular figures, or are repelled and recede from one another.[4]

And toward the end of his *Opticks* he added to a long series of "Queries" about the results of corpuscular action:

All these things being consider'd, it seems probable to me, that God in the Beginning form'd Matter in solid, massy, hard, impenetrable, moveable Particles, of such Sizes and Figures, and with such other Properties, and in such Proportion to Space, as most conduced to the End for which he form'd them. . . . And therefore, that Nature may be lasting, the Changes of corporeal Things are to be placed only in the various Separations and new Associations and Motions of these permanent Particles. . . . It seems to me farther, that these Particles have not only a *Vis inertiae* [inertial Force], accompanied with such passive Laws of Motion as naturally result from that Force, but also that they are moved by certain active Principles, such as is that of Gravity, and that which causes [chemical] Fermentation, and the Cohesion of Bodies.[5]

These statements and others like them describe the Newtonianism that played so large a role in the thought of the eighteenth and nineteenth centuries. After Newton's death in 1727 most scientists and educated laymen conceived the universe to be an infinite neutral space inhabited by an infinite number of corpuscles whose motions were governed by a few passive laws like inertia and by a few active principles like gravity. From these premises Newton had deduced with unprecedented precision most of the known phenomena of optics and all the known phenomena of celestial and terrestrial mechanics, including both the tides and the precession of the equinoxes. Beginning where he had left off, his successors tried to discover the additional force laws required to account for the remaining natural phenomena: heat, electricity, magnetism, cohesion, and, above all, chemical combination. At last the crumbling Aristotelian universe was replaced by a comprehensive and coherent world-view, and a new chapter in man's developing conception of nature was begun.

The New Fabric of Thought

The construction of Newton's corpuscular world machine completes the conceptual revolution that Copernicus had initiated a century and a half earlier. Within this new universe the questions raised by Copernicus' astronomical innovation were at last resolved, and Copernican astronomy became for the first time physically and cosmologically plausible. The earth's relation to the other bodies in the universe was once again defined. Men again knew why a shot fired into space would return to the point from which it had departed, though they now understood that the shot must not be fired quite vertically. Only as Copernicanism became credible, through the dissemination and acceptance of this new conceptual framework, did the last significant opposition to the conception of a planetary earth disappear. Newton's universe was not, however, merely a framework for Copernicus' planetary earth. Far more important, it was a new way of looking at nature, man, and God — a new scientific and cosmologic perspective which, during the eighteenth and nineteenth centuries, repeatedly enriched the sciences and reshaped both religious and political philosophy.

The same Newtonian principles which, by providing an economical derivation and a plausible explanation of Kepler's Laws, closed the astronomical revolution also supplied astronomy itself with a host of powerful new research techniques. For example, when improved quantitative techniques of telescopic observation showed that the planets do not really quite obey Kepler's Laws, Newtonian physics made it possible first to explain and then to predict the planets' minor deviations from their fundamental elliptical orbits. As Newton's derivation had shown, Kepler's Laws should apply rigorously only if the sun exerts the sole attractive force on each planet. But planets also attract each other, particularly when they approach and pass, and this extra attraction draws them out of their fundamental orbits and changes their speeds. During the eighteenth century mathematical extensions of Newton's work enabled astronomers to predict these deviations with great precision, and during the nineteenth century the inversion of this predictive technique was responsible for one of astronomy's greatest triumphs. In 1846 Leverrier in France and Adams in England independently predicted the existence and the orbit of a previously unsuspected planet which they believed was the cause of unexplained irregularities in the orbit of the known planet Uranus. When the telescope was turned to the heavens, the new planet, Neptune, was discovered, dimly visible, within a degree of the position predicted by Newtonian theory.

Examples of Newtonianism's fruitfulness in astronomy could be multiplied almost endlessly, and astronomy was not the only science affected. To examine but one of the possible examples from one of many sciences, consider the impact of Newton's work upon chemical experimentation during the eighteenth century. In spite of his explicit intention Newton led most of his successors to believe that gravity, and therefore weight, were intrinsic properties of matter. He thus gave weight a new significance in science. It became, for the first time, an unequivocal measure of quantity of matter, and as a result the balance became a fundamental chemical instrument. The balance alone could tell the chemist how much matter was put into a chemical reaction and how much was evolved. Since antiquity chemists had believed that quantity of matter was conserved during chemical reactions, but there had been no widely accredited measure of "quantity of matter." In the climate of Aristotelian or even Cartesian thought weight was usually regarded, like color, texture, or hardness, as a secondary characteristic of matter — one that might be altered by the process of chemical reaction. The concept of weight as a universally accredited tool for "balancing" chemical reactions and for determining whether matter is lost to or gained from an unsuspected source during such reaction was therefore a partial product of Newtonianism. That new tool was one of the several important roots of the revolution in chemical thought that centered about the Frenchman Lavoisier during the last decades of the eighteenth century.

A whole book would be required to transform and multiply these two isolated examples — the discovery of Neptune and the new significance of weight — into a balanced discussion of the effects of the new universe upon science, and the discussion would still be incomplete. In the conceptual fabric that grew up around the new universe nonscientific thought was transformed as well. In the infinite and multipopulated universe conceived by seventeenth-century scientists and philosophers the localization of heaven in the skies and of hell beneath the earth's crust became mere metaphors, dying echoes of a symbolism that had once had concrete geographic significance. Simultaneously the conception of a universe constructed of atoms which move forever in accordance with a few God-given laws changed many men's image of the Deity Himself. In the clockwork universe God frequently appeared to be only the clockmaker, the Being who had shaped the atomic parts, established the laws of their motion, set them to work, and then left them to run themselves. Deism, an elaborated version of this view, was an important ingredient in late seventeenth- and eighteenth-century thought. As it advanced, the belief in miracles declined, for miracles were a suspension of mechanical law and a direct intervention by God and his angels in terrestrial affairs. By the end of the eighteenth century an increasing number of men, scientists and nonscientists alike, saw no need to posit the existence of God.

Other reflections of the new science can be discovered in the political philosophy of the eighteenth and nineteenth centuries. Several recent writers have pointed to the significant parallels between the seventeenth-century conception of a mechanically functioning solar system and the eighteenth-century conception of a smoothly running society. The system of checks and balances incorporated in the Constitution of the United States, for example, was intended to give the new American society the same sort of stability in the presence of disruptive forces that the exact compensation of inertial forces and gravitational attraction had given to the Newtonian solar system. Also, the eighteenth century's determination to derive the characteristics of a good society from the innate characteristics of the individual man may well have been fostered in part by the corpuscularism of the seventeenth century. In eighteenth- and nineteenth-century thought the individual appears again and again as the atom from which the mechanism, society, is fabricated. In the opening paragraphs of the Declaration of Independence, Jefferson derived the right to revolution from the God-given or inalienable rights of the social atom, man, and his derivation seems to parallel the one in which Newton, a century earlier, had derived the mechanism of nature from the God-given or innate properties of the individual physical atom.

Even these few disparate and undeveloped examples indicate that with the creation of the Newtonian universe our story has come full circle. What the Aristotelian universe had done for earth-centered astronomy the Newtonian universe was to do for Copernican astronomy. Each was a world view that tied astronomy to other sciences and related it to nonscientific thought as well; each was a conceptual tool, a way of organizing knowledge, evaluating it, and gaining more; and each dominated the science and philosophy of an age. Having traversed this circle from world view to world view, we may at last realize the sense in which it turns upon Copernicus' astronomical innovation. The conception of a planetary earth was the first successful break with a constitutive element of the ancient world view. Though intended solely as an astronomical reform, it had destructive consequences which could be resolved only within a new fabric of thought. Copernicus himself did not supply that fabric; his own conception of the universe was closer to Aristotle's than to Newton's. But the new problems and suggestions that derived from his innovation are the most prominent landmarks in the development of the new universe which that innovation had itself called forth. The creation of the need and the aid supplied in its fulfillment are the contributions to history that constitute the Copernican Revolution.

Its historical contributions do not, however, exhaust the Revolution's significance. Because it illustrates the continuing cyclic process by which knowledge grows, the Copernican Revolution has a larger importance as well. The last two and one-half centuries have proved that the conception of the universe which emerged from the Revolution was a far more powerful intellectual tool than the universe of Aristotle and Ptolemy. The scientific cosmology evolved by seventeenth-century scientists and the concepts of space, force, and matter that underlay it, accounted for both celestial and terrestrial motions with a precision undreamed of in antiquity. In addition, they guided many novel and immensely fruitful research programs, disclosing a host of previously unsuspected natural phenomena and revealing order in fields of experience that had been intractable to men governed by the ancient world view. These are permanent achievements. As long as the continuous tradition of Western learning survives, scientists will be able to explain the phenomena first elucidated by Newtonian concepts, just as Newton was able to explain the more restricted list of phenomena previously elucidated by Aristotle and Ptolemy. That is how science advances: each new conceptual scheme embraces the phenomena explained by its predecessors and adds to them.

But though the achievements of Copernicus and Newton are permanent, the concepts that made those achievements possible are not. Only the list of explicable phenomena grows; there is no similar cumulative process for the explanations themselves. As science progresses, its concepts are repeatedly destroyed and replaced, and today Newtonian concepts seem no exception. Like Aristotelianism before it, Newtonianism at last evolved — this time within physics — problems and research techniques which could not be reconciled with the world view that produced them. For half a century we have been in the midst of the resulting conceptual revolution, a revolution that is once again changing the scientist's (though not yet the layman's) conception of space, matter, force, and the structure of the universe. Because they provide an economical summary of a vast quantity of information, Newtonian concepts are still in use. But increasingly they are employed for their economy alone, just as the ancient two-sphere universe is employed by the modern navigator and surveyor. They are still a useful aid to memory, but they are ceasing to provide a trustworthy guide to the unknown.

Therefore, though more powerful than its predecessors, the Newtonian universe is not proving more final. Nor is its history, considered as one of many chapters in the development of human thought, very different in structure from that of the earth-centered universe which Copernicus and Newton destroyed. This book is one long chapter in a continued and continuing story.

TECHNICAL APPENDIX

1. Correcting Solar Time

In the central chapters of this book we have assumed that if the apparent solar day is defined as the time interval between one local noon and the next, then the time required by the stars to complete their diurnal circles is invariably just 4 minutes (more accurately 3 minutes 56 seconds) shorter than this solar day. But as noted in a footnote to Chapter 1, this is not quite the case. If the intervals between successive local noons are perfectly regular, then the stars must move at an irregular rate. Conversely, if the stars complete successive diurnal circles in equal intervals of time, then the lengths of successive solar days must vary. This fact was recognized in antiquity, at least by the time of Ptolemy and very probably before. To understand it let us assume, as the ancients did, that the apparent motion of the stars is perfectly regular, so that the stars provide a fundamental time scale. We shall then discover two distinct reasons why the intervals between the instants when the sun achieves its maximum daily elevation must vary.

The first cause of the irregularity of apparent solar time is the variation in the rate at which the sun seems to move through the zodiacal constellations. As we discovered in Chapter 2, the sun moves more rapidly along the ecliptic from autumnal to vernal equinox than from vernal to autumnal. In its daily race with the stars the sun therefore seems to lose ground more rapidly in winter than in summer, so that if time is measured by the stars, the sun must take longer to regain maximum elevation during the winter than it requires in summertime. It follows that the apparent solar day should be longest in midwinter and shortest in midsummer, and this would be the case if another cause of irregularity did not intervene.

The second source of the apparent solar day's variability is the angle at which the ecliptic intersects the equator on the celestial sphere. To understand its effect look again at Figure 13, Chapter 1, and imagine that equally spaced lines of celestial longitude are drawn on the sphere, just as lines of longitude are drawn on any terrestrial globe. For the sake of simplicity, assume in addition that the sun's apparent motion along the ecliptic is perfectly regular and at the rate of 1° along the great circle per day. It then turns out that, because the ecliptic is tilted with respect to the equator, the net *eastward* motion of the sun varies from one day to the next. When the sun is at or near one of the solstices, its apparent motion with respect to the stars is very nearly parallel to the celestial equator. In addition, it is moving on a part of the sphere where the lines of longitude are somewhat closer together than they are at the equator. As a result the net eastward motion of the sun is somewhat more than 1° *of celestial longitude* per day, and the celestial sphere must therefore turn westward through slightly more than 361° in order to carry the sun from maximum elevation to maximum elevation. At the equinoxes the situation is quite different. There the lines of celestial longitude have their maximum spacing on the sphere. Furthermore, the sun's constant total motion is to the northeast or southeast rather than due east, and it therefore does not move eastward as much as 1° a day. As a result the celestial sphere need not rotate through quite 361° to return the sun to maximum elevation. This effect, considered alone, makes the apparent solar day longest at the solstices and shortest at the equinoxes.

In order to correct for these two irregularities modern civilizations have adopted a time scale known as mean solar time, whose fundamental time unit is the *average* length of the apparent solar day. On this time scale the stars do, by definition, move perfectly regularly, completing their diurnal circles in just 23 hours 56 minutes 4.091 seconds. But the scale that makes the stars regular makes the sun irregular. For example, the sun's maximum elevation rarely occurs at local noon, mean solar time. The time kept by sundials, the only instruments that directly measure apparent solar time, does not pass at the same rate as the time kept by our watches or announced by time signals on the radio. During December and January, when both the effects discussed above act to shorten the apparent solar day, the interval between successive maximum elevations of the sun is very nearly 0.5 minute less than the mean solar day. Furthermore, the effect of this small discrepancy is cumulative — apparent time runs slower than mean time for many days in a row — so that at one season of the year the sun reaches maximum elevation (apparent noon) almost 20 minutes before mean solar noon. At other seasons apparent time runs faster than mean time, and over the years the two stay together. But they are rarely together during any one day. In order to keep accurate time by the sun it is therefore necessary to correct the sundial by using a table, or a diagram like the one shown in Figure 53.

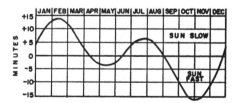

Figure 53. A graph of the equation of time, indicating the annual variation of the difference between mean and apparent solar time.

The preceding discussion of time employs the apparent motion of the stars as its standard of regularity. Clearly this choice of standard is arbitrary, at least from a logical viewpoint. Logically we might equally well have chosen the sun's apparent motion as our standard of regularity and shown that on the corresponding time scale the stars move at a continually varying rate. But the choice of the sun as a standard of regularity would be immensely inconvenient to both science and civil life. The diagram of Figure 53 would have to be applied to clocks and watches rather than to sundials. Astronomers and physicists would be forced to describe the earth's axial rotation as occurring at a constantly varying rate. The stellar standard avoids these awkwardnesses. It is well adapted to all civil and most scientific functions.

Yet it has not turned out to be quite adequate for science, at least not for scientific theory; the time scale implicit in Newton's Laws of Motion does not quite correspond to the stellar standard. From Newton's Laws, as they are now understood, it is possible to show that the earth's axial rotation is being gradually slowed by effects like tidal friction and that, as a result, the apparent stellar motions are very gradually slowing down. Either the Laws or the stellar standard must therefore be adjusted, and considerations of scientific convenience suggest the search for a new standard. To date the theoretical inadequacy of the stellar standard is without practical significance. But it has an immense importance to science, and it has therefore led scientists to a renewed search, which continues actively today, for a clock that will conform to the time scale of scientific theory more accurately than the celestial machine itself.

2. Precession of the Equinoxes

A second technical simplification introduced in the body of this book·was the neglect of the precession of the equinoxes. This is the effect, mentioned briefly in Chapter 1, that results in a slow motion of the celestial pole through the stars. If we had been concerned only with naked-eye observations made during the lifetime of a single astronomer, our simplification would have been entirely appropriate — naked-eye observations cannot disclose its inaccuracy unless they are made at widely separated points in time. But observations made, for example, two centuries apart show that, while the stars themselves retain constant relative positions, the celestial pole about which they move gradually shifts its position among them at a rate just over 0.5° per century. Observations repeated over far longer periods disclose the pattern of this precessional motion; as the centuries pass the pole of the heavens moves gradually through the stars in a circle, completing one revolution every 26,000 years. The center of this circle is the pole of the ecliptic — the point at which a diameter perpendicular to the plane of the ecliptic intersects the celestial sphere — and the radius of the circle is just 23½°, the same as the angle in which the celestial equator intersects the ecliptic on the sphere of the stars (Figure 54a).

The precessional motion seems to have been noticed first by the Hellenistic astronomer Hipparchus during the second century B.C., and, though not
widely known at first, it was discussed by a number of subsequent astronomers, including Ptolemy. Most of Ptolemy's Moslem successors described
some form of the effect, and by adding a ninth sphere to the ancient system
they succeeded in explaining it physically. Their most popular explanation
is indicated diagramatically in Figure 54b, which shows the three outermost
spheres of the set; N and S are the north and south celestial poles, and the
exterior sphere rotates westward about them, once every 23 hours 56
minutes, as the sphere of the stars had rotated in the older system. The next

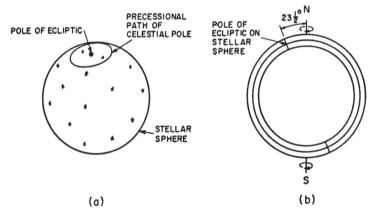

(a)   (b)

Figure 54. The precession of the equinoxes. Diagram (a) shows the circle on
the celestial sphere around which the celestial pole moves once in every 26,000
years. The center of the circle is the pole of the ecliptic, and all points on the circle
are just 23½° from this center. Diagram (b) shows how the Moslems explained
precession with the aid of a ninth sphere, the outer sphere in the drawing. This
ninth sphere rotates once every 23 hours 56 minutes, as the sphere of the stars had
rotated in the older eight-sphere systems. The eighth sphere, on which the stars are
set, rotates about its own poles once in 26,000 years, thus slowly changing the
position of the celestial pole among the stars. Inside the eighth sphere is the
sphere of Saturn, which encloses the remaining planetary spheres as in the older
systems.

sphere, the middle one in the diagram, is the sphere that carries the stars,
and it is joined to the outermost sphere by an axis which passes through the
poles of the ecliptic on the sphere of the stars and through two points 23½°
from the poles on the outer sphere. This new sphere of the stars is whirled
around daily by the outermost sphere (this accounts for the diurnal stellar
circles). In addition it has a slow motion of its own, one rotation every
26,000 years, which gradually changes the relations between the individual

stars and the celestial poles. The innermost of the three spheres is Saturn's, and it is drawn as a thick shell to allow space for the epicyclic components of Saturn's motion. By itself this thick sphere, connected to the sphere of the stars by an axis through the poles of the ecliptic, accounts for Saturn's average circular motion through the stars.

In the context of ancient and medieval astronomical thought this ninth-sphere explanation of precession seems both simple and natural. In fact, it compares relatively well with the Copernican explanation — a gradual conical motion of the earth's axis which, during the course of 26,000 years, is directed successively to all the points on a circle of radius 23½° about the pole of the ecliptic. Until Newton explained precession as a physical consequence of the moon's gravitational attraction for the equatorial bulge of the earth, both Copernican and Ptolemaic astronomers required one extra and physically superfluous motion in order to account for it.° Precession has, therefore, no logical bearing upon the transition from an earth-centered to a sun-centered universe.

Historically, however, the problem of explaining precession had a significant role in inaugurating the Copernican Revolution. It helped to make Ptolemaic astronomy seem monstrous. The observational consequences of precession are very small even when observations extend over several centuries, and a small error in the data can therefore result in a radical change in the description of the over-all phenomenon. Both Hipparchus and Ptolemy described precession in a way qualitatively equivalent to the one represented by Figure 54, but many of their contemporaries denied the existence of the effect entirely or else described it quite differently. Particularly in the Moslem world a number of divergent descriptions of precession were prevalent. There was no agreement about its rate — in fact, many astronomers believed that the rate varied. In addition, there was an important school which believed that even the direction of precession changed periodically, an effect known as trepidation. Brahe's observations were required before astronomers could again recognize the true simplicity of the phenomenon. Copernicus himself did not improve the situation in the slightest. He added extra circles to his system in order to account for the gradual change in the precessional rate and for other nonexistent effects. But though Copernicus did not improve the account of precession given by ancient and medieval astronomers, he was immensely concerned to do so, and that concern provided an important impetus to astronomical reform. In

---

° Copernicus himself did not require an extra motion to account for precession, because he had already introduced one in another connection. He used an annual conical motion to keep the earth's axis parallel to itself throughout the year (Figure 31b), and he could therefore explain precession by giving this conical motion a period very very slightly less than a year. But Copernicus' successors, who thought that a single orbital motion would keep the earth's axis perpetually in alignment, did need an additional conical motion with a period of 26,000 years in order to explain the changing position of the celestial pole.

Copernicus' day an adequate account of precession was the principal prerequisite for the most pressing problem of practical astronomy, the reform of the Julian calendar.

To discover the effect of precession upon the design of calendars, return once more to Figure 54. As the diagram shows, the position of the ecliptic upon the sphere of the stars is fixed once and for all. But though the changing positions of the celestial poles do not affect the ecliptic, they do change the position of the celestial equator and therefore of the equinoxes, the points at which the ecliptic and the celestial equator intersect. During the precessional period, 26,000 years, each of the equinoxes moves slowly and steadily around the ecliptic at the rate of about 1½° per century. Therefore, the length of time required by the sun to move once around the ecliptic (the so-called sidereal year) is not quite the same as the length of time it requires to move on the ecliptic from vernal equinox to vernal equinox (the tropical year). The latter, which is more than 20 minutes shorter than the former, is vastly more difficult to measure, because it refers the sun's motion to an imaginary and moving point rather than to a fixed star. But the tropical year is the year of the seasons, and it is this that must be measured with precision before an accurate long-term calendar can be designed. Copernicus' concern with the calendar therefore led him to a serious study of precession, and thus to an intimate knowledge of that aspect of astronomy about which Ptolemaic astronomers were in the greatest disagreement. It is the problem of precession which underlies Copernicus' remark that "the mathematicians . . . cannot even explain or observe the constant length of the seasonal year" (p. 137), and it is this remark which heads his list of motives for innovation.

3. Phases of the Moon and Eclipses

Because it is identical with the modern explanation, the ancients' account of the cause of the moon's phases played no role in the Copernican Revolution, and it could therefore be omitted from the earlier chapters of this book. But the phases of the moon play a direct role in the ancient measurements of the dimensions of the universe, and these measurements, as we have repeatedly noted, helped make the ancient two-sphere universe seem concrete and real to scientists and nonscientists alike. Besides, the ancient explanation of phases, as well as the correlated explanation of eclipses, provides an important additional illustration of the scientific adequacy of the ancient world view.

The explanation with which we are concerned was well known in Greece by the end of the fourth century B.C., though it may have originated considerably earlier. With the acceptance of the two-sphere universe came the larger and less well documented assumption that all the celestial wanderers were spheres as well. In part this assumption derived from analogy to the spherical shape of the earth and heavens, and in part from the conception of the perfection of the spherical shape and its appropriateness to the perfect heavens. More direct, though incomplete, evidence was provided by the observed cross sections of the sun and moon. Now if the moon is a sphere, a distant sun can illuminate only one-half of its surface (Figure 55a), and the fraction of this illuminated hemisphere visible to an observer will necessarily vary with his position. An observer on the sun would see the entire hemisphere at all times; an observer on the earth looking toward the moon when it lay between him and the sun would see none of the illuminated hemisphere whatsoever. It follows that the portion of the moon's surface clearly visible to a terrestrial observer must depend upon the relative positions of the sun, the moon, and the earth.

Four *relative* positions of the sun and moon at four equally spaced periods during the lunar month are shown in Figure 55b, which portrays the earth-centered orbits of the sun and moon in the plane of the ecliptic. (Since only relative positions are significant in discussions of the moon's phases, the diagram can readily be adapted to a sun-centered universe.)

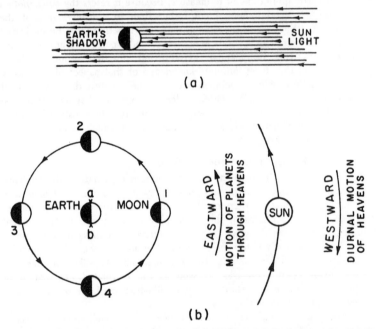

Figure 55. The ancient (and modern) explanation of the moon's phases. Diagram (*a*) indicates that only half of the surface of a sphere is illuminated by the rays of the distant sun. Diagram (*b*) shows the portion of this illuminated hemisphere visible to a terrestrial observer for various *relative* positions of the sun, earth, and moon. Position 1 is new moon; 2 is the waxing half moon; 3, full moon; and 4, the waning half moon.

A westward rotation of the entire diagram, excepting the central earth, accounts for the diurnal motion of the sun and moon, so that an observer at *a* sees the sun just setting and one at *b* sees it rising. Only the eastward orbital motions of the sun and moon are motions with respect to the diagram. When the moon is at position 1 in the diagram it rises with the sun, but, since its dark side is pointed toward the earth, it can scarcely be seen by a terrestrial observer. This is the position of new moon. Slightly more than a week later the moon's rapid orbital motion has carried it 90° east of the slow-moving sun where it appears, relative to the sun, in position 2. It now rises at noon and is near the zenith at sunset. Only half of the disk is clearly visible from the earth, so that this is the position of first quarter. After another week or a bit more, the moon is full and rises as the sun sets (position 3). Third quarter is shown at position 4, corresponding to a moon that rises around midnight and is near the zenith at sunrise.

The diagram used in deciphering the moon's phases can also be used in the explanation of eclipses: as the moon moves from position 2 to position 4, it may pass through the earth's shadow, in which case it grows dim and is eclipsed. If the moon always appeared on the ecliptic, it would be eclipsed each time it reached position 3, but, since it continually wanders north and south, the full moon, earth, and sun rarely lie on a straight line. Full moon must lie close to the ecliptic for a lunar eclipse, and this cannot happen more than twice a year and seldom happens that often. Solar eclipses occur whenever the moon, at position 1, casts its shadow on the earth, and this happens relatively frequently, at least twice a year. Yet solar eclipses are rarely seen by terrestrial observers. The moon's shadow on the earth is extremely small, and an observer must be in the shadow to see the eclipse. Besides, the moon rarely blocks off more than a small fraction of the sun's disk. Therefore, an observer at any one location can seldom see even a partial eclipse of the sun and may never see a total one. For him it will be a rare, striking, and sometimes terrifying phenomenon.

4. Ancient Measurements of the Universe

One of the most interesting technical applications of ancient astronomy was its use in the determination of cosmological distances and sizes which could not be measured directly, that is, by ordinary measuring sticks. These distance measurements illustrate the world view's fruitfulness with greater immediacy than most of its other applications, because the mathematical operations upon which they depend lose all physical significance unless certain essential elements in the conceptual scheme are true. For example, whether the earth is a flat disk or a sphere, the stars do appear to move in diurnal circles, and techniques that describe this apparent motion are therefore useful whatever their conceptual basis. But only if the earth is really a sphere can it be said to have a circumference that can be determined from the observations of the skies discussed below.

The first reference to measurements of the earth's circumference appears in Aristotle's writings, so that such measurements were probably made by the middle of the fourth century B.C. But we know only the results of these earliest measurements, not the method employed; the first measurement of which we have a relatively complete, though second-hand, account is the one made by Eratosthenes, the librarian of the great manuscript collection at Alexandria, during the third century B.C. Eratosthenes measured the angle *a* (Figure 56) between the rays of the noon sun and a vertical

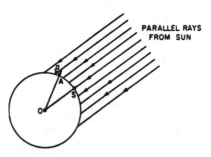

Figure 56. Eratosthenes' measurement of the earth's circumference. If *S* is due south of *A* on the earth's surface, then the distance *AS* must be the same fraction of the earth's circumference as the angle *a* is of 360°.

gnomon located at Alexandria, *A*, on a day when the noon sun was directly overhead at Syene, *S*, a second Egyptian city located 5000 stades due south of Alexandria. This angle he found to be 1/50 of a full circle (or 7 1/5°). Since all the rays striking the surface of the earth from the very distant sun may be considered parallel, the angle *a*, which is the sun's distance from the zenith at Alexandria, is equal to the angle *AOS* subtended by *S* and *A* at the center of the earth, *O*. Furthermore, since this angle is just 1/50 of a circle, the distance from Alexandria to Syene must be 1/50 of the circumference of the earth, and the total circumference must be 50 × (distance from Alexandria to Syene) = 50 × 5000 = 250,000 stades. Most modern students believe that Eratosthenes' figure is approximately 5 percent lower than the result given by modern measurement (24,000 miles), but unfortunately it is impossible to be sure. The length of the unit "stade" used by Eratosthenes is unknown, and the known location of Alexandria and Syene cannot be used to define the unit, because both the "5000" and the "1/50" used in the computation above have clearly been "rounded off" to make the report easier to read.

A second group of measurements was made during the second century B.C. by Aristarchus of Samos, now more famous for his anticipation of the Copernican system. He estimated the distance to and the sizes of the sun and moon in terms of the angle *MES* subtended by the centers of the sun and moon at the earth when the moon is exactly half full (Figure 57). Since the moon can be half full only if the angle *EMS* subtended by the earth and the sun at the moon is exactly a right angle, the size of *MES* must completely determine the shape of the right triangle whose vertices are the moon, the earth, and the sun. Aristarchus' measurement gave *MES* = 87°, which corresponded to a triangle in which *ES*:*EM*::19:1. Ac-

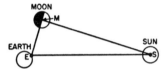

Figure 57. Aristarchus' measurement of the relative distances from the earth to the moon and the sun. When the moon is exactly half full, the angle *EMS* must be just 90°. Therefore a measurement of the angle *MES* will determine the ratio of *EM* to *ES*, that is, the ratio of the moon's distance from the earth to that of the sun.

cordingly, he reported that the sun was 19 times as far from the earth as the moon and that, since the moon and the sun subtend the same angle at the earth (Figure 58), it was also 19 times as large.

Modern measurements, made by quite different techniques and with the aid of telescopes, indicate that Aristarchus' ratio was too small by a factor of more than twenty; the ratio *ES*:*EM* is very nearly 400:1, not 19:1. This discrepancy arises from the measurement of the angle *MES* which should be 89° 51′, rather than 87°. In practice that measurement is very difficult, particularly with the instruments known to have been available to Aristarchus. The precise centers of the sun and moon are very hard to determine; in addition it is difficult to be sure when the moon is *just* half full. Given these problems, Aristarchus seems to have chosen the smallest angle compatible with his uncertain observations, presumably in order to keep the resulting ratio credible. Similar considerations must have motivated his successors, for,

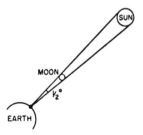

Figure 58. The large but distant sun and the smaller but nearby moon subtend the same angle at the earth's surface.

though appreciably improved, estimates of the relative distances to the sun and the moon remained too small throughout antiquity and the Middle Ages.

The preceding measurements yield only the ratios of astronomical distances, but by an immensely ingenious argument Aristarchus was able to convert them to absolute magnitudes, that is, he was able to determine the diameter of and distances to the sun and the moon in stades. His results were derived from observations of a lunar eclipse of maximum duration, an eclipse during which the moon lies squarely 'on the ecliptic and therefore passes through the very center of the earth's shadow. First he measured the time that elapsed between the instant when the edge of the moon first entered the shadow and the instant when the moon was totally obscured for the first time. This figure he compared with the length of time during which the moon was totally obscured, and he thus discovered that the period of total obscurity was approximately the same length as the period required for the moon to enter into the earth's shadow. He concluded that the breadth of the earth's shadow in the region where it is crossed by the moon is very nearly twice the diameter of the moon itself.

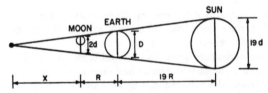

Figure 59. Aristarchus' construction for computing the absolute distances to the moon and sun in terms of the observations made during a lunar eclipse.

Figure 59 shows the astronomical configuration which Aristarchus analyzed. In the diagram the moon is shown immediately after it has fully entered the shadow of the earth. The diameter of the moon is d (an unknown) and the diameter of the earth's shadow at the moon is therefore $2d$; the diameter of the earth is D (known in stades from Eratosthenes' measurement of the circumference of the earth); and the distance from the moon to the earth is R (again an unknown to be determined). Finally, since the sun's diameter and distance from the earth are just 19 times the moon's, the diameter of the sun's disk is just $19d$ and its distance from the earth is just $19R$. So Aristarchus' problem, and ours, is to determine d and R, the unknown distances, in terms of the earth's diameter, D, a quantity whose value in stades has already been determined.

The diagram shows immediately three similar triangles whose bases are of length $2d$, D, and $19d$, and whose altitudes are respectively x (an unknown), $x + R$, and $x + 20R$. (Actually the bases of the three triangles are very slightly shorter than the diameters with which they have been equated above, but if the triangles are extremely acute, as they are, this discrepancy is too small to affect the result.) The ratio of the altitude to the base of the smallest triangle must be the same as that of the largest, or,

$$\frac{x}{2d} = \frac{x + 20R}{19d}.$$

Multiplying both sides of the equation by $38d$ yields a new equation: $19x = 2x + 40R$, so that $x = 40R/17$. In other words, the earth's shadow extends beyond the moon for a distance about 2⅓ times the distance from the earth to the moon.

Comparing the smallest of the triangles with the triangle of intermediate size gives another equation, from which d may be determined. The first comparison gives:

$$\frac{x}{2d} = \frac{x + R}{D}.$$

Substituting $40R/17$ for x and multiplying both sides by $17/R$ yields:

$$\frac{20}{d} = \frac{40 + 17}{D}.$$

From the last equation, $d = 20D/57 = 0.35D$. That is, the diameter of the moon is just greater than one-third the diameter of the earth, and since the sun's diameter is just 19 times the moon's, the sun must have just over 6⅔ times the diameter of the earth.

Since D, the diameter of the earth, is known, the actual sizes of the sun and moon are given by the computation above. Their distances can be determined by a small additional computation. Because both the sun and the moon subtend an angle of 0.5° at the earth, each could be placed 720 times on the circumference of a full (360°) circle with its center at the earth. The distance of the moon from the earth must therefore be the radius of a circle whose circumference is 720 times the moon's diameter, now known, and the sun's distance must be just 19 times as great. Since the circumference of a circle is 2π times its radius, the moon's distance from the earth must be just over 40 diameters of the earth and the distance to the sun should be approximately 764 earth diameters.

The methods employed in these computations are brilliant, typifying the very best efforts of Greek scientists, but the numerical results, particularly those concerning the sun, are uniformly inaccurate because of the initial error in the determination of the angular separation of the sun and the half moon. Modern measurements give the moon's diameter as just over one-fourth the diameter of the earth and its distance as approximately 30 earth diameters, neither of which is far from the values computed by Aristarchus. But the sun's diameter is now thought to be almost 110 times that of the earth and the distance to the sun is about 12,000 earth diameters, both very much larger than Aristarchus supposed. Though various corrections to Aristarchus' measurements were made during antiquity and though the possibility of significant error in the measured distance to the sun was often recognized, all ancient and medieval estimates of this cosmological dimension remained vastly too small.

Because it depends only upon the relative positions of earth, moon, and sun, Aristarchus' techniques for determining size and distance can be applied with equal accuracy or inaccuracy in the Ptolemaic and Copernican universes. The ancient determinations of astronomical dimensions could, therefore, have no direct role in the Copernican Revolution. But they did have several indirect ones, all of which helped to strengthen the Ptolemaic system. The possibility of making astronomical measurements illustrated the great fruitfulness of the Aristotelian-Ptolemaic universe. In addition, the results of the measurements helped to make the ancient cosmology seem real by increasing the concreteness with which its structure was specified. Finally, and most important, the measurement of the distance to the moon provided an astronomical yardstick which, during the Middle Ages, was used to provide an indirect measure of the size of the entire universe.

As indicated early in Chapter 3, medieval cosmologists often supposed that each crystalline shell was just thick enough to contain the epicycle of its planet and that the shells as a group nested so that they filled all of space. Using these hypotheses mathematical astronomers were able to determine the relative sizes and thicknesses of all the shells. These relative dimensions were then converted to earth diameters, stades, or miles, by using Aristarchus' method of determining the distance to the moon's sphere. A typical set of the cosmological dimensions that resulted was included in the original discussion. It indicates the detail with which the universe was investigated and understood by pre-Copernican scientists.

REFERENCES

Chapter 1. The Ancient Two-Sphere Universe

1. Sir Thomas L. Heath, *Greek Astronomy*, Library of Greek Thought (London: Dent, 1932), pp. 5–7.

2. Benjamin Jowett, *The Dialogues of Plato*, 3rd ed. (London: Oxford University Press, 1892), III, 452–453.

Chapter 2. The Problem of the Planets

1. Vitruvius, *The Ten Books on Architecture*, trans. M. H. Morgan (Cambridge: Harvard University Press, 1926), pp. 261–262.

2. Sir Thomas L. Heath, *Aristarchus of Samos* (Oxford: Clarendon Press, 1913), p. 140.

Chapter 3. The Two-Sphere Universe in Aristotelian Thought

1. Aristotle, *On the Heavens*, trans. W. K. C. Guthrie, The Loeb Classical Library (Cambridge: Harvard University Press, 1939), p. 91 (279a6–17).

2. *Ibid.*, pp. 243–253 (296b8–298a13).

3. Sir Thomas L. Heath, *Greek Astronomy*, Library of Greek Thought (London: Dent, 1932), pp. 147–148.

4. Aristotle, *On the Heavens*, p. 345 (310b2–5).

5. Aristotle, *Physics*, trans. P. H. Wickstead and F. M. Cornford, The Loeb Classical Library (Cambridge: Harvard University Press, 1929), I, 331 (213a31–34).

6. Aristotle, *On the Heavens*, pp. 23–25 (270b1–24).

7. Jean Piaget, *The Child's Conception of Physical Causality*, trans. Marjorie Gabain (London: Kegan Paul, Trench, Trubner, 1930), pp. 110–111.

8. Heinz Werner, *Comparative Psychology of Mental Development*, rev. ed. (Chicago: Follett, 1948), pp. 171–172.

9. Aristotle, *Physica*, trans. R. P. Hardie and R. K. Gaye, in *The Works of Aristotle*, II (Oxford: Clarendon Press, 1930), 208b8–22.

Chapter 4. Recasting the Tradition: Aristotle to the Copernicans

1. St. Augustine, *Works*, ed. Marcus Dods (Edinburgh: Clark, 1871–77), IX, 180–181.

2. St. Thomas Aquinas, *Commentaria in libros Aristotelis De caelo et mundo*, in *Sancti Thomae Aquinatis . . . Opera Omnia*, III (Rome: S. C. de Propaganda Fide, 1886), p. 24. My translation.

3. St. Thomas Aquinas, *The "Summa Theologica,"* Part I, Questions L–LXXIV, trans. Fathers of the English Dominican Province, 2nd ed. (London: Burns Oates & Washbourne, 1922), p. 225 (Q. 68, Art. 3).

4. Aquinas, *Summa Theologica,* Part III, Questions XXVII–LIX (London: Washbourne, 1914), pp. 425, 433 (Q. 57, Arts. 1, 4). Quoted by permission of the publishers and of Benziger Brothers, Inc., New York.

5. Charles H. Grandgent, *Discourses on Dante* (Cambridge: Harvard University Press, 1924), p. 93.

6. Dante, *The Banquet,* trans. Katharine Hillard (London: Routledge and Kegan Paul, 1889), pp. 65–66.

7. *Ibid.,* pp. 69, 79–80.

8. Nicole Oresme, *Le livre du ciel et du monde,* ed. A. D. Menut and A. J. Denomy, in *Mediaeval Studies,* III–V (Toronto: Pontifical Institute of Mediaeval Studies, 1941–43), IV, 243.

9. *Ibid.,* IV, 272. In translating this and the following passages from Oresme's commentary I have often been guided by the longer English selection in the mimeographed pamphlet, "Selections in Medieval Mechanics," prepared by Marshall Clagett of the University of Wisconsin and kindly made available to me by its author. Professor Clagett's translations from Oresme and many other medieval scientific writers will soon appear under the title *Mechanics in the Middle Ages.*

10. *Ibid.,* p. 273.

11. Condensed with permission from Marshall Clagett's "Selections in Medieval Mechanics" (see n. 9), pp. 35–39. The original is Jean Buridan, *Quaestiones super octo libros physicorum* (Paris, 1509), Book VIII, Question 12. I have introduced a few purely stylistic changes and dropped one set of italics.

12. Clagett, "Selections," p. 40, from Buridan, *Quaestiones.*

13. *Mediaeval Studies,* IV, 171.

14. Alfred North Whitehead, *Science and the Modern World* (New York: Macmillan, 1925), p. 19.

15. Quoted by John Herman Randall, Jr., *The Making of the Modern Mind,* 2nd ed. (Boston: Houghton Mifflin, 1940), p. 213.

16. Sir Thomas L. Heath, *A History of Greek Mathematics* (Oxford: Clarendon Press, 1921), I, 284.

17. Quoted by Edward W. Strong, *Procedures and Metaphysics* (Berkeley: University of California Press, 1936), p. 43, from Thomas Taylor, *The Philosophical and Mathematical Commentaries of Proclus on the First Book of Euclid's Elements* (London, I [1788] and II [1789]).

18. Marsilio Ficino, *Liber de Sole,* in *Marsilii Ficini Florentini, . . . Opera* (Basel: Henric Petrina, [1576]), I, 966. My translation.

19. Quoted and translated by Edwin A. Burtt, *The Metaphysical Foundations of Modern Physical Science,* 2nd ed. (New York: Harcourt, Brace, 1932), p. 48, from a fragment of one of Kepler's early disputations.

Chapter 5. Copernicus' Innovation

1. All the quotations in Chapter 5 are from the Preface and Book I of Copernicus' *De Revolutionibus Orbium Caelestium* (1543). The translation was prepared by John F. Dobson and Selig Brodetsky and published as *Occasional Notes of the Royal Astronomical Society,* vol. 2, no. 10 (London: Royal Astronomical Society, 1947). In reprinting the translation I have consistently replaced the word "orbit" by either "sphere" or "circle" (see Edward Rosen, *Three Copernican Treatises* [New York: Columbia University Press, 1939], pp. 13–16, for the difficulties inherent in Copernicus' use of the Latin *orbis*). In perhaps a dozen other

places I have suppressed other examples of modern terminology or have made similar minor changes for the sake of increased clarity. In making such changes I have repeatedly been guided by the very useful Latin-French edition of Copernicus' Book I prepared from the authoritative Thorn edition (1873) by Alexandre Koyré (Paris: Félix Alcan, 1934). I am indebted to the Royal Astronomical Society for permission to reprint so large a portion of their translation.

Chapter 6. The Assimilation of Copernican Astronomy

1. Quoted by Francis R. Johnson, *Astronomical Thought in Renaissance England* (Baltimore: Johns Hopkins Press, 1937), p. 207. I have modernized spelling and punctuation in this and many of the other quotations in this chapter.

2. *Ibid.*, pp. 188–189, from the translation (1605) by Joshua Sylvester.

3. Translated and quoted by Dorothy Stimson, *The Gradual Acceptance of the Copernican Theory of the Universe* (New York, 1917), pp. 46–47, from Bodin's *Universae Naturae Theatrum* (Frankfort, 1597).

4. Translated and quoted by Andrew D. White, *A History of the Warfare of Science with Theology in Christendom* (New York: Appleton, 1896), I, 126.

5. *Ibid.*, pp. 126–127, from Melanchthon's *Initia Doctrinae Physicae*.

6. *Ibid.*, p. 127.

7. John Donne, "Ignatius, his Conclave," in *Complete Poetry and Selected Prose of John Donne*, ed. John Hayward (Bloomsbury: Nonesuch Press, 1929), p. 365.

8. *Ibid.*, p. 202.

9. John Milton, *Paradise Lost*, Book I, line 26.

10. Nicole Oresme, *Le livre du ciel et du monde*, in *Mediaeval Studies*, IV, 276.

11. Translated and quoted by James Brodrick, *The Life and Work of Blessed Robert Francis Cardinal Bellarmine*, *S.J.* (London: Burns Oates and Washbourne, 1928), II, 359.

Chapter 7. The New Universe

1. Robert Hooke, *An Attempt to Prove the Motion of the Earth from Observations* (London: John Martyn, 1674), reproduced in R. T. Gunther, *Early Science in Oxford*, VIII (Oxford: privately printed, 1931), pp. 27–28.

2. Newton, *Opticks*, 4th (1730) ed. (New York: Dover, 1952), p. 401.

3. *Ibid.*

4. Newton, *Mathematical Principles of Natural Philosophy*, ed. Florian Cajori (Berkeley: University of California Press, 1946), p. xviii.

5. Newton, *Opticks*, pp. 400–401.

BIBLIOGRAPHICAL NOTES

Introductory

The following notes provide both an indication of my most essential debts to previous studies and a convenient entrance to the vast labyrinth of scholarly literature dealing with the history of astronomy and related fields before 1700. Where possible I have restricted the discussion to books readily available in English. Articles, monographs, and studies in foreign languages are, with few exceptions, cited only if they have contributed substantially to my own account of the Copernican Revolution or if they are omitted (as many recent studies are) from the principal bibliographical sources below. A few minor sources, included in the list of references, are omitted below.

Fuller bibliographies for several portions of the field will be found in M. R. Cohen and I. E. Drabkin, *A Source Book of Greek Science* (New York, 1948); E. J. Dijksterhuis, *De Mechanisering van het Wereldbeeld* (Amsterdam, 1950); F. Russo, *Histoire des sciences et des techniques: bibliographie* (Paris, 1954); and George Sarton, *A Guide to the History of Science* (Waltham, Mass., 1952). Exhaustive bibliographies for several relevant topics will be found in George Sarton, *Introduction to the History of Science,* 3 vols. in 5 (Baltimore, 1927–1947) and in the annual bibliographies in the journal *Isis.* Many of the books cited in other connections below also contain useful bibliographic information. The more recent works, particularly A. C. Crombie, *Augustine to Galileo* (Cambridge, Mass., 1952), and A. R. Hall, *The Scientific Revolution, 1500–1800* (London, 1954), will prove especially useful.

All the general histories of science discuss the period and many of the problems covered by this book, but only Herbert Butterfield, *The Origins of Modern Science, 1300–1800* (London, 1949), has had particular influence on the structure of this book. Marshall Clagett, *Greek Science in Antiquity* (New York, 1955) and Hall, *Scientific Revolution* (above), give extremely useful background for their respective periods, though neither was available to me until my manuscript was in substantially final form. Dijksterhuis, *Mechanisering* (above), is also an essential source for those who can read Dutch.

Bertrand Russell, *A History of Western Philosophy* (New York, 1945), and W. Windelband, *A History of Philosophy*, trans. J. H. Tufts (New York, 1901), have provided useful background on the development of philosophy. J. L. E. Dreyer, *A History of Astronomy from Thales to Kepler*, 2nd ed. (New York, 1953); Lynn Thorndike, *A History of Magic and Experimental Science*, 6 vols. (New York, 1923–1941); and Sarton, *Introduction* (above), have been consulted so often in the composition of this book that I cite them, below, only for those parts of it that follow their treatments closely. Pierre Duhem, *Le système du monde*, 6 vols. (Paris, 1913–1954), which could be used in the same way, I have consulted for special topics.

Chapters 1 and 2

R. H. Baker, *Astronomy*, 5th ed. (New York, 1950) is an excellent source for the requisite information about technical astronomy.

George Sarton, *A History of Science: Ancient Science through the Golden Age of Greece* (Cambridge, Mass., 1952), surveys Egyptian, Mesopotamian, and Hellenic astronomy within the context of ancient science and culture. O. Neugebauer, *The Exact Sciences in Antiquity* (Princeton, 1952), provides a more detailed introduction to Egyptian and Babylonian astronomy from their beginning through the Hellenistic period, though its selectivity may mislead some readers about the important role played by the Hellenic astronomical tradition. Neugebauer's second edition (Providence, R.I., 1957) includes in an appendix a very useful description of the astronomical devices elaborated in Ptolemy's *Almagest*. Sir Thomas L. Heath, *Aristarchus of Samos* (Oxford, 1913), is the standard source for Greek astronomy to the third century B.C. Chapters 7–9 of Dreyer, *History* (above), discuss Greek astronomy from Apollonius through Ptolemy.

Many selections from ancient astronomical writings can be found in Sir Thomas L. Heath, *Greek Astronomy* (London, 1932), and in Cohen and Drabkin, *Source Book* (above). Other relevant passages are Plato, *Timaeus*, in *The Dialogues of Plato*, ed. Benjamin Jowett, 3rd ed. (London, 1892), III, and Vitruvius, *The Ten Books on Architecture*, trans. M. H. Morgan (Cambridge, Mass., 1926), though the translation of the latter is occasionally handicapped by ignorance of astronomical fact and theory. Ptolemy, *The Almagest*, has recently been translated by R. Catesby Taliaferro in *Great Books of the Western World*, vol. XVI (Chicago, 1952). Detailed studies will, however, continue to depend upon the standard edition, *Syntaxis mathematica*, ed. J. L. Heiberg, 2 vols. (Leipzig, 1898–1903).

Much information about ancient calendars is contained in many of the secondary sources above. More detailed studies are F. H. Colson, *The Week* (Cambridge, Mass., 1926), and R. A. Parker, *The Calendars of Ancient Egypt* (Chicago, 1950). Stonehenge as a primitive observatory is discussed in Sir Norman Lockyer, *Stonehenge and Other British Stone Monuments Astronomically Considered*, 2nd ed. (London, 1909), but see also Jacquetta Hawkes, "Stonehenge," *Scientific American* CLXXXVIII (June 1953), 25–

31. For the role of the heavens in primitive cosmological thought see Henri Frankfort *et al.*, *The Intellectual Adventure of Ancient Man* (Chicago, 1946), and Heinz Werner, *The Comparative Psychology of Mental Development*, rev. ed. (Chicago, 1948).

Chapter 3

The principal sources for this chapter are Aristotle's writings on the physical sciences, particularly his *Physics, Metaphysics, On the Heavens, Meteorology*, and *On Generation and Corruption*. All are translated in *The Works of Aristotle Translated into English*, ed. Sir William David Ross, 12 vols. (Oxford, 1928–1952) and all except the last in the editions of The Loeb Classical Library. The annotations and text in Sir W. D. Ross, *Aristotle's Physics* (Oxford, 1936), make it a particularly useful edition.

John Burnet, *Early Greek Philosophy*, 3rd ed. (London, 1920); Theodor Gomperz, *Greek Thinkers*, trans. Laurie Magnus (I) and G. A. Berry (II–IV) (New York, 1901–1912); and Kathleen Freeman, *The Pre-Socratic Philosophers* (Oxford, 1946) make it possible to set Aristotle's thought in the tradition established by his predecessors. Sir W. D. Ross, *Aristotle*, 3rd ed. (London, 1937) and Werner Jaeger, *Aristotle: Fundamentals of the History of His Development*, trans. Richard Robinson (Oxford, 1934) are important biographical studies of the works. F. M. Cornford, *The Laws of Motion in Ancient Thought* (Cambridge, Eng., 1931), deals incisively with a number of special problems treated in this chapter.

The post-Ptolemaic computations of cosmological dimensions from the principle of plenitude are discussed by Edward Rosen, "A Full Universe," *Scientific Monthly* LXIII (1946), 213–217, and in Chapters VIII and XI of Dreyer, *History* (above, Introductory section). The evidence for the experiment at Pisa is analyzed by Lane Cooper, *Aristotle, Galileo, and the Leaning Tower of Pisa* (Ithaca, 1935), but Cooper's work should be supplemented by the discussions of the development of Galileo's laws cited below for Chapters 4 and 7. Primitive conceptions of space and motion are discussed by Werner (above, Chapter 1) and in the numerous works of Jean Piaget, particularly *The Child's Conception of the World*, trans. Joan and Andrew Tomlinson (London, 1929); *The Child's Conception of Physical Causality*, trans. Marjorie Gabain (London, 1930); and *Les notions de mouvement et de vitesse chez l'enfant* (Paris, 1946).

Chapter 4

The most salient aspects of the transition from Hellenic to Hellenistic science are sketched in George Sarton, *Ancient Science and Modern Civilization* (Lincoln, Neb., 1954). Far more detail is provided in the same author's *Introduction* (above, Introductory section).

Henry Osborn Taylor, *The Mediaeval Mind*, 4th ed., 2 vols. (Cambridge, Mass., 1925) discusses the early Christian depreciation of pagan science, and Dreyer, *History* (above, Introductory), provides much relevant illustration for astronomy. Important primary sources are Augustine, *Confessions* and *Enchiridion*, in *Works*, ed. Marcus Dods (Edinburgh, 1871–1877).

My account of the reconciliation of Aristotelian cosmology with Biblical history derives from St. Thomas Aquinas, *The "Summa Theologica,"* trans. Brothers of the English Dominican Province, 22 vols. (London, 1913–1925) and from the *Commentaria* on Aristotle's physical treatises in vols. II and III of St. Thomas Aquinas, *Opera Omnia*, 12 vols. (Rome, 1882–1906). The result of the new integration is illustrated in Dante, *The Banquet*, trans. Katharine Hillard (London, 1889), and *The Divine Comedy* (many English editions). The effects of the cosmological metaphor on medieval and Renaissance thought are sketched in Charles H. Grandgent, *Discourses on Dante* (Cambridge, Mass., 1924), and in S. L. Bethell, *The Cultural Revolution of the Seventeenth Century* (London, 1951).

Arabic and medieval European astronomy are treated by Dreyer, *History*; by Duhem, *Le système*; by Sarton, *Introduction* (all above, Introductory); and by Lynn Thorndike, *Science and Thought in the Fifteenth Century* (New York, 1929). Thorndike feels that previous scholars have dated the emergence of an erudite European astronomical tradition too late, but, at least with respect to the problem of the planets, I find his evidence unconvincing. Additional valuable information is contained in Derek J. Price, ed., *Equatorie of the Planetis* (Cambridge, England, 1955).

A. C. Crombie, *Augustine to Galileo* (above, Introductory), gives the best substantive and bibliographic survey of medieval science. My own discussion is also indebted to numerous special studies, particularly Carl Boyer, *The Concepts of the Calculus*, 2nd ed. (Wakefield, Mass., 1949); Marshall Clagett, *Giovanni Marliani and Late Medieval Physics* (New York, 1941), and "Some General Aspects of Physics in the Middle Ages," *Isis* XXXIX (1948), 29–44; Alexandre Koyré, *Études Galiléennes*, 3 vols. (Paris, 1939); Annaliese Maier, *Studien zur Naturphilosophie der Spätscholastik*, 4 vols. (Rome, 1951–1955); and John Herman Randall, Jr., "The Development of Scientific Method in the School of Padua," *Journal of the History of Ideas* I (1940), 177–206. Koyré and Randall give particularly useful illustrations of the transmission of scholastic ideas to the early founders of modern science. Primary sources for the study of scholastic theories of motion include Thomas Bradwardine, *Tractatus de Proportionibus*, ed. and trans. H. Lamar Crosby, Jr. (Madison, Wis., 1955); Marshall Clagett, "Selections in Medieval Mechanics" (Madison, Wis., mimeographed, no date); Jean Buridan, *Quaestiones super libris quattuor de caelo et mundo*, ed. Ernest A. Moody (Cambridge, Mass.: Mediaeval Academy of America, 1942); and Nicole Oresme, *Le livre du ciel et du monde*, ed. A. D. Menut and A. J. Denomy, *Mediaeval Studies* III–V (Toronto, 1941–1943).

The relation between science and a variety of social, economic, and intellectual currents of the Renaissance is discussed by John Herman Randall, Jr., *The Making of the Modern Mind*, rev. ed. (Boston, 1940), and by Myron P. Gilmore, *The World of Humanism, 1453–1517* (New York, 1952). Ancient and Renaissance Neoplatonism are discussed in Lynn Thorndike, *Magic and Experimental Science* (above, Introductory), and Arthur O. Lovejoy, *The Great Chain of Being* (Cambridge, Mass., 1948). Henry Osborn Taylor, *Thought and Expression in the Sixteenth Century*, 2 vols. (New York, 1920) includes a description of Renaissance Neoplatonism. Plato's attitude toward mathematics is treated by Sir Thomas L. Heath, *A History of Greek Mathematics*, 2 vols. (Oxford, 1921), and the effects of this attitude, in its Neoplatonic form, on science are discussed from a variety of points of view in Edwin Arthur Burtt, *The Metaphysical Foundations of Modern Physical Science* (New York, 1932); Alexandre Koyré, "Galileo and Plato," *Journal of the History of Ideas* IV (1943), 400–428; and Edward W. Strong, *Procedures and Metaphysics* (Berkeley, Calif., 1936). The last work is the only one that gives adequate emphasis to the mystical and ascientific tenor of Neoplatonic thought, but it may go too far in concluding that no point of view so fundamentally irrational could have had a fruitful effect on the practice of science. For Neoplatonism see also the works relating to Nicholas of Cusa and Giordano Bruno cited below for Chapter 6.

Chapter 5

Copernicus' life and work are well described in Angus Armitage, *Copernicus, The Founder of Modern Astronomy* (London, 1938), but this account should be supplemented by the far fuller one in Ludwig Prowe, *Nicolaus Coppernicus*, 2 vols. (Berlin, 1883–1884). Copernicus' minor astronomical works and Rheticus' *Narratio prima* are translated with an excellent introduction and notes in Edward Rosen, *Three Copernican Treatises* (New York, 1939). The only complete English translation of Copernicus' principal work is Nicolaus Copernicus, *On the Revolutions of the Heavenly Spheres*, trans. Charles Glenn Wallis, in *Great Books of the Western World*, vol. XVI (Chicago, 1952). Anyone using this edition should first consult the highly critical review by O. Neugebauer in *Isis* XLVI (1955), 69–71. A useful English translation of the Preface and Book I of the *De Revolutionibus* by John F. Dobson and Selig Brodetsky has been published as *Occasional Notes of the Royal Astronomical Society*, vol. II, no. 10 (London, 1947). Alexandre Koyré, *Copernic: Des Révolutions des Orbes Célestes* (Paris, 1934), provides a convenient French and Latin edition of Book I with all the prefatory materials as well as a penetrating and provocative introductory discussion. The standard edition of the complete text is Maximilian Curtze, *Nicolai Copernici Thorunensis: De revolutionibus orbium caelestium libri VI* (Torun, 1873). Important special aspects of Copernicus' astronomy are discussed by Dreyer, *History* (above, Introductory), and of his physics and cosmology by Edgar Zilsel, "Copernicus and Mechanics," *Journal of the History of Ideas* I (1940), 113–118.

Chapter 6

Much useful material on the sixteenth- and seventeenth-century reactions to Copernican astronomy is contained in Francis Johnson, *Astronomical Thought in Renaissance England* (Baltimore, 1937); Grant McColley, "An early friend of the Copernican theory: Gemma Frisius," *Isis* XXVI (1937), 322–325; Dorothy Stimson, *The Gradual Acceptance of the Copernican Theory of the Universe* (New York, 1917); Lynn Thorndike, *Magic and Experimental Science* (above, Introductory), particularly vol. V, chap. 18, and vol. VI, chaps. 31 and 32; and Andrew D. White, *A History of the Warfare of Science with Theology in Christendom*, 2 vols. (New York, 1896). Thorndike's material is the richest and most balanced, but it must be used cautiously because it is occasionally predicated on important elementary errors about the technical relations between Copernican and Ptolemaic astronomy (see, for example, the sentence connecting pp. 424 and 425 in vol. V).

The fullest and most recent account of Galileo's conflict with the Church is Giorgio de Santillana, *The Crime of Galileo* (Chicago, 1955). Some of the older discussions are, however, still useful, particularly Karl von Gebler, *Galileo Galilei and the Roman Curia*, trans. Mrs. George Sturge (London, 1879), and James Brodrick, S. J., *The Life and Work of Blessed Robert Francis Cardinal Bellarmine*, 2 vols. (London, 1928).

On Tycho Brahe see J. L. E. Dreyer, *Tycho Brahe* (Edinburgh, 1890), and also his *Opera Omnia*, ed. J. L. E. Dreyer, 15 vols. (Hauniae, 1913–1929). The often underestimated popularity of the Tychonic system has been effectively documented by Grant McColley in "Nicolas Reymers and the Fourth System of the World," *Popular Astronomy* XLVI (1938), 25–31, and "The Astronomy of Paradise Lost," *Studies in Philology* XXXIV (1937), 209–247.

There is no adequate English study of Kepler's life or work, but Carola Baumgardt, *Johannes Kepler: Life and Letters* (New York, 1951), includes some useful quotations from sources. The standard study, Max Caspar, *Johannes Kepler* (Stuttgart, 1948), should soon be available in translation. The relevant works must still be read in the fine *Gesammelte werke*, ed. Max Caspar, 12 vols. (Munich, 1938–1955). R. H. Baker, *Astronomy* (above, Chapter 1), gives a technical account of Kepler's Laws from a modern viewpoint. Much information about their technical development is included in Dreyer, *History* (above, Introductory), and in A. Wolf, *A History of Science, Technology, and Philosophy in the XVI and XVII Centuries*, rev. ed. prepared by Douglas McKie (London, 1950). Other important studies of Kepler are cited for Chapter 7, below.

Galileo's telescopic observations are discussed in many of the preceding studies. They are, however, best approached directly in Galileo Galilei, *The Sidereal Messenger*, trans. Edward Stafford Carlos (London, 1880), and *Dialogue on the Great World Systems*, ed. Giorgio de Santillana (Chicago, 1953). Some indication of the telescope's immense impact on the scientific and popular imagination is provided in Marjorie Hope Nicolson, "A World in the Moon," *Smith College Studies in Modern Languages* XVII, no. 2 (Northampton, Mass., 1936); in Martha Ornstein, *The Role of Scientific Societies in the Seventeenth Century* (Chicago, 1938); in some selections from *The Portable Elizabethan Reader*, ed. Hiram Haydn (New York, 1946); and in Edward Rosen, *The Naming of the Telescope* (New York, 1947). Most of Galileo's work is beyond the scope of this book; some other important studies are, however, cited for Chapters 4 and 7.

Chapter 7

Pre-Copernican and post-Copernican ideas about the infinity of the universe are discussed in Francis R. Johnson and Sanford V. Larkey, "Thomas Digges, the Copernican System, and the Idea of the Infinity of the Universe," *Huntington Library Bulletin* V (April 1934), 69–117; Alexandre Koyré, "Le vide et l'espace infini au XIV siècle," *Archives d'histoire doctrinale et littéraire du moyen âge* XXIV (1949), 45–91; Lovejoy, *Great Chain* (above, Chapter 4); Grant McColley, "Nicolas Copernicus and an Infinite Universe," *Popular Astronomy* XLIV (1936), 525–533, and "The Seventeenth-Century Doctrine of a Plurality of Worlds," *Annals of Science* I (1936), 385–430. McColley's articles are particularly informative, though he seriously overstates the case for Copernicus' belief in an infinite universe. Johnson's article reprints the relevant portions of Digges's *Perfit Description*. Other useful primary sources are Nicholas of Cusa, *Of Learned Ignorance*, trans. Germain Heron (London, 1950), and excerpts from *De ludo globi* chosen and translated by Maurice de Gandillac in *Oeuvres choisies de Nicolas de Cues* (Paris, 1942). See also the annotated translation of Bruno's *On the Infinite Universe and Worlds* in Dorothea Waley Singer, *Giordano Bruno, His Life and Thought* (New York, 1950). For some time to come the standard source on this problem will surely be Alexandre Koyré, *From the Closed World to the Infinite Universe* (Baltimore, 1957).

Despite the wealth and range of this literature, there still seems to be one important gap in our knowledge of the evolution of the concept of an infinite Copernican universe. From Bruno's death in 1600 to the publication of Descartes's *Principles of Philosophy* in 1644, no Copernican of any prominence appears to have espoused the infinite universe, at least in public. After Descartes, however, no Copernican seems to have opposed the conception. This silence during the first half of the seventeenth century is understandable, but it leaves a puzzle about the development and propagation of the belief in a physically infinite universe.

Frederick A. Lange, *The History of Materialism*, trans. E. C. Thomas, 3rd ed. (New York, 1950), and Kurd Lasswitz, *Geschichte der Atomistik*, 2nd ed., 2 vols. (Hamburg, 1926), include much essential information about the development of atomism since antiquity. Seventeenth-century atomism is thoroughly surveyed in Marie Boas, "The Establishment of the Mechanical Philosophy," *Osiris* X (1952), 412–541, a monograph which also provides a fine bibliography on this topic. Significant special studies of atomism's role in the development of modern science include Fulton H. Anderson, *The Philosophy of Francis Bacon* (Chicago, 1948); Marie Boas, "Boyle as a Theoretical Scientist," *Isis* XLI (1950), 261–268; Thomas S. Kuhn, "Robert Boyle and Structural Chemistry in the Seventeenth Century," *Isis* XLIII (1952), 12–36; and Paul Mouy, *Le développement de la physique Cartésienne* (Paris, 1934). Important and representative primary sources for the central tenets of this seventeenth-century tradition are René Descartes, *Les principes de la philosophie* and *Le monde ou le traité de la lumière* in vols. IX and XI of the *Oeuvres de Descartes*, ed. Charles Adam and Paul Tannery (Paris, 1904 and 1909), and Robert Boyle, *Origin of Qualities and Forms*, in vol. II of *The Works*, ed. A. Millar (London, 1744).

For the problems presented to terrestrial physicists by the Copernican theory see Alexandre Koyré, *Études Galiléennes*, 3 vols. (Paris, 1939), "Galileo and the Scientific Revolution of the Seventeenth Century," *Philosophical Review* LII (1943), 333–348, and especially "A Documentary History of the Problem of Fall from Kepler to Newton," *Transactions of the American Philosophical Society* (n.s.) XXXXV, no. 4 (1955), 329–395. Kepler's celestial mechanics is discussed in Dreyer, *History* (above, Introductory); in Gerald Holton, "Johannes Kepler's Universe: Its Physics and Metaphysics," *American Journal of Physics*, XXIV (1956), 340–351; and in Alexandre Koyré, "La gravitation universelle, de Kepler à Newton," *Archives internationales d'histoire des sciences* XXX (1951), 638–653. Borelli's system is described in Angus Armitage, " 'Borelli's Hypothesis' and the Rise of Celestial Mechanics," *Annals of Science* VI (1950), 268–282, and Alexandre Koyré, "La méchanique céleste de J. A. Borelli," *Revue d'histoire des sciences* V (1952), 101–138. The work of Robert Hooke is treated in its relation to Newton by Louise D. Patterson, "Hooke's Gravitation Theory and its Influence on Newton," *Isis* XL (1949), 327–341, and XLI (1950), 32–45, and more incisively with the aid of a new document by Alexandre Koyré, "An Unpublished Letter of Robert Hooke to Isaac Newton," *Isis* XLIII (1952), 312–337. Many documents illustrating Hooke's work will be found in R. T. Gunther, *Early Science in Oxford*, 14 vols. (Oxford, 1920–1945), particularly vols. VI and VIII.

Guidance to the vast literature on Newton can be found in almost all the bibliographical sources listed in the Introductory section, above. Whatever is novel in my own approach concerns Newton's atomism and the related metaphysical substructure of the *Principia*. This analysis is at least partially derived from and supported by Florian Cajori, "Ce que Newton doit à Descartes," *L'enseignement mathématique* XXV (1926), 7–11, and "Newton's Twenty Years' Delay in Announcing the Law of Gravitation," in *Sir Isaac Newton*, ed. History of Science Society (Baltimore, 1928); A. R. Hall, "Sir Isaac Newton's Note-Book, 1661–65," *Cambridge Historical Journal* IX (1948), 239–250; Alexandre Koyré, "The Significance of the Newtonian Synthesis," *Archives internationales d'histoire des sciences* XXIX (1950), 291–311; Thomas S. Kuhn, "Newton's '31st Query' and the Degradation of Gold," *Isis* XLII (1951), 296–298, and "Preface to Newton's Optical Papers," in I. B. Cohen, ed., *Isaac Newton's Letters and Papers on Natural Philosophy* (Cambridge, Mass., in press), and S. I. Vavilov, "Newton and the Atomic Theory," in *The Royal Society Newton Tercentenary Celebrations* (Cambridge, Eng., 1947). Convenient primary sources are Sir Isaac Newton, *Mathematical Principles of Natural Philosophy*, ed. Florian Cajori (Berkeley, Calif., 1946), and *Opticks*, reprint ed. (New York, 1952).

Technical Appendix

R. H. Baker, *Astronomy* (above, Chapter 1), discusses the equation of time, precession of the equinoxes, eclipses, and the moon's phases from a modern viewpoint. Heath, *Aristarchus* (above, Chapter 1), and Dreyer, *History* (above, Introductory) have much historical information on all these subjects except the first, on which see A. Rome, "Le problème de l'equation du temps chez Ptolémée," *Annales de la société scientifique de Bruxelles* (Series 1) LIX (1939), 211–224. Heath and Dreyer also treat the ancient determinations of astronomical dimensions, on which see also Aubrey Diller, "The Ancient Measurements of the Earth," *Isis* XL (1949), 6–12. Additional details about the Moslem complications of the treatments of precession will be found in Francis J. Carmody, *Al-Bitrûjî: de motibus celorum* (Berkeley, Calif., 1952), and "Notes on the Astronomical Works of Thâbit b. Qurra," *Isis* XLVI (1955), 235–242.

INDEX

The more important page references are in italic type

Absolute space. *See* Aristotelian space

Adams, J. C., 261

Aether, 79–82, 91–92

Al Fargani, 81–82, 160. *See also* Distance, astronomical

Alpetragius, 177

Amici, G. B., 138

Anaximander, 26–27

Anima motrix, 214–216, 230, *245–249*

Aphelion, 214–215

Apogee, 147

Apollonius, 59, 72–74

Aquinas, Thomas, 109–111

Aristarchus, *42,* 144, 160, *274–278*

Aristotelian cosmology, 59, 79–84, 91–92, 109, 112. *See also* Cosmology, Christian

Aristotelian space, 79, 87–88, 97–98, 231

Aristotelian theory of motion, 82, 84–87, 95–98, 118–119, 151, 244

Aristotle, 78, 105, 111–112; criticism, 83, 106, 109, 115–123, 151–153, 206, 208; authority, 78, 83–84, 95–99, 103, 111, 115, 117, 127

Astrology, 25, 49–50, 92–94, 98, 107, 206; relation to astronomy, 94

Astronomical tables, 103, 105, *187–188,* 196, 219

Astronomy, prehistoric, *9–11, 13–14,* 47, 49–50; Hellenic, 5, *26–27,* 50, 52, 104–105; Hellenistic, 66, 72–73, 104–105; Moslem, 72, 74, 102, 112, 269–270; medieval, 74, 112–113, 123–124; Renaissance, 67, 125–126. *See also* Cosmology

Atomism, *41,* 199, 230, 232, *235–237.* *See also* Corpuscular philosophy

Augustine, 107, 128

Bellarmine, Cardinal, *198,* 226

Blundeville, Thomas, 186

Bodin, Jean, 190

Bonamico, F., 120

Borelli, G. A., *248–249,* 253–254

Brahe, Tycho, 93, 211, 221; role in Copernican Revolution, *201,* 204–206; observations, 160, *200–201,* 212–214. *See also* Tychonic system

Bruno, Giordano, *199,* 220, *235–237,* 243

Buridan, Jean, 119–122, 124

Calendars, *8–12,* 46–47, 125–126, 196, 271

Callippus, *58,* 80

Calvin, 125, *192,* 195–196

Capella, Martianus, 178

Catholic Church, attitude toward astronomy, 93, 106–112, 197; and Copernicanism, 106, 192, *195–199,* 226. *See also* Copernican cosmology

Celestial mechanisms, 27, 52, 79–80, 89, 105, 114, 121–122, *214–216,* 240–242, 245–252, *254–260*

Celestial sphere. *See* Sphere of stars

Celestial-terrestrial dichotomy, 43, 91–92, 94, 121–122, 153, 207–208, 221–222, 237, 252. *See also* Sublunary region; Superlunary region

Collisions, atomic, *238–240,* 242

Columbus, Christopher. *See* Voyages

Comets, 45, 93, *208,* 244

Compass points, 9

Conceptual schemes, commitment to, 39, 75–77, 83; economy, *36–37,* 54, 74, 77, 90, 169, *212–213,* 226, 265; fruitfulness, *39–41,* 54, 64–66, 77, 208–209, 212–213, 261–262, 264, 273, 278; distinguished from observations, *25–26,* 35; psychological functions, *38–39,* 105, 113. *See also* Scientific revolutions

Conjunction, *49,* 178

Constellations, 13–16; Big Dipper, *13–16,* 20–21, 206; zodiacal, *25,* 46, 93, 266

Copernican astronomy, 135, 143–144, 149, 155–172, 179–180, *210–214,* 270n; construed as computing device, *187–188,* 196, 226; harmonies, 142, 172–181, 184; compared with Ptolemaic, 75, 162–163, 165–166, 169–175, 178–179, 181–182, 223; reception, 131, 172, 182–183, *185–188,* 201, 209–210, 224, 227–228; compared with Tychonic, 202–205. *See also* Copernican cosmology

Copernican cosmology, 7, 40–41, 94, 145–155, 220, 224, *231–232;* anticipations, *41–42,* 115–116, 142, 144, 149, *233–237;* and physics, 86–87, 115–123, 150–155, 214–215, *230–231,* 238, 243–245, 252–254, 261–

LINCOLN CHRISTIAN COLLEGE AND SEMINARY